Solid-State Servicing

by

William Sloot

HOWARD W. SAMS & CO., INC.
THE BOBBS-MERRILL CO., INC.
INDIANAPOLIS · KANSAS CITY · NEW YORK

FIRST EDITION

THIRD PRINTING—1976

International Standard Book Number: 0-672-20888-1
Library of Congress Catalog Card Number: 72-182874

Preface

The purpose of this book is to provide the service technician with information which will help him service solid-state electronic equipment, primarily of the home-entertainment variety. Although a great deal of material has been published over the last twenty years explaining transistor operation, many technicians remain confused when attempting to service solid-state circuits. This may be due primarily to the attempted analogy between vacuum-tube and transistor operation. While transistor operation and vacuum-tube operation are quite similar regarding the processing of signals throughout the circuit, the dynamics of the processing are quite different. Rather than clarifying a point, these analogies often imply functions which force erroneous conclusions. Therefore, these analogies will be avoided in this book.

The intent of this book is to present a new approach or format to bridge the gap between useful theory and practical servicing. The book is divided into two parts. In Part 1, certain statements will be made regarding solid-state electronics. These statements will then be explained, with emphasis on how they apply to the troubleshooting of electronics. Each statement is intended to be a practical bit of knowledge which can be directly applied to the solution of real service problems. A certain amount of overlap and interdependence is inevitable, and all of the knowledge gained is, to a degree, supplemental. Some of the statements in Part 1 are the ground rules which provide a foundation, or common reference, for adequate communication. Finally, at the end of certain sections, questions will be asked concerning problems which can be solved by application of the principles stated.

Part 2 illustrates a cross section of circuits used in home-entertainment products. Typical circuits are usually represented rather than specific circuits, since the majority of manufacturers provide detailed information and schematics. The primary intent of this section is to provide troubleshooting hints and procedures which will enable the technician to rapidly isolate most problems. A thorough knowledge of circuit operation is usually not necessary to successfully repair the majority of home-entertainment products. For this reason, lengthy circuit descriptions will be avoided in this book.

Electronics is a rapidly changing science. Therefore, the circuitry presented in this book is described in terms of function rather than specifics. Although circuits and devices change, basic signal processing changes very little. Thinking in terms of function and stressing signal processing rather than method allows application of troubleshooting techniques which do not rapidly become obsolete. Radio and tv are not segregated into separate sections, since radio circuits have their counterpart in television. For example, television tuners and radio tuners perform the same function even though the operating frequencies are different.

Test equipment is specified for the various service techniques. The best instrument for a specific problem depends on the circumstance and the ability of the technician. An experienced technician can often accomplish more with a jumper wire than an inexperienced man with an oscilloscope. Therefore, in the final analysis, the best equipment must be the brain of the technician. No amount of test equipment can substitute for logic and experience.

Questions are posed at the end of certain sections. However, these questions are designed to illustrate a point and should not be considered a testing procedure.

Certain prerequisites are assumed throughout the book. These requirements are a knowledge of basic electronics and some prior experience servicing electronic equipment.

WILLIAM SLOOT

Contents

PART I

SOLID STATEMENTS

A. Electron Flow Is From Negative to Positive

The direction in which electrons or current are considered to flow through a device or circuit is of little importance. However, one direction or the other must be chosen to analyze or discuss circuit operation. The above statement provides the ground rules for the direction current or electron flow will be considered to have in this book. The direction was selected in preference over the so-called "conventional" current flow because most electronics schools teach the negative-to-positive concept. Also, this concept conforms more closely to electronic theory concerning device operation. When a direction of current flow has been established, a number of useful applications can result.

Transistors are power amplifying devices that are current controlled (i.e., they draw output current proportional to the input current). For this reason, understanding and troubleshooting transistor circuits is more easily accomplished by thinking in terms of current paths, rather than concentrating on potentials existing at various points within the circuit. Transistor circuits contain two current paths: the input or control (base/emitter) circuit and the output or controlled (emitter/collector) circuit.

Applications

1. The correct polarity of voltage drop across a component can be determined by knowing the direction of electron flow. The voltage drop across a device is such that the end where electrons enter is negative (−) and where they exit is positive (+). However, this does not apply to a device acting as the power source, such as a battery.

2. The correct operating polarities for a transistor can be easily determined. Simply polarize in a manner to cause current flow against the emitter arrow in both the input and output circuits. Likewise, the current flows against the arrow (the anode) in a diode.

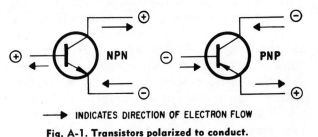

INDICATES DIRECTION OF ELECTRON FLOW

Fig. A-1. Transistors polarized to conduct.

Transistor operation depends on current flow. If the polarities of the applied voltages are not as illustrated in Fig. A-1, current will not flow and the transistor will not operate. The concept of establishing current paths in this manner makes circuit analysis very easy and is the format that generally will be followed in this book. Notice that the polarities applied to npn and pnp transistors must be exactly opposite to establish current flow. This is the only fundamental difference between these two transistor types that need concern the service technician. Most low-power silicon transistors used in home-entertainment products are of the npn variety. Therefore, most of the illustrations and examples in this book are npn transistors.

3. The transistor type can be readily determined if the basing is known. Polarize the transistor to cause current flow. The polarity will then indicate the transistor type. Connect one lead of an ohmmeter (the polarity of the ohmmeter leads must be known beforehand) to the base and the other lead to either the collector or emitter. Place the ohmmeter function switch in the R × 100 position. If a reading is obtained with the positive (+) lead connected to the base, the device is an npn. If a reading is obtained with the negative (−) lead connected to the base, the device is a pnp. The middle letter of the device type indicates the polarity. An npn transistor has a positive base and a pnp has a negative base.

B. A Transistor Can Be Represented as Two Diodes
Connected Back to Back

Resistance measurements between the three transistor leads would seem to indicate that a transistor is two diodes connected back to back, as shown in Fig. B-1. This is indeed the case, as far as simple resistance measurements are concerned. Actual transistor operation cannot be explained in this manner, since two diodes connected as illustrated will not amplify. The concept does, however, have certain useful applications.

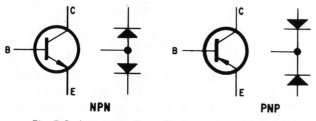

Fig. B-1. A transistor is analogous to a pair of diodes connected back to back.

Applications

1. A transistor can be quickly checked for major failures with an ohmmeter. Simply check the base-emitter and base-collector junctions for good diode action (very high resistance in one direction and low resistance in the other). Also check the resistance between the emitter and the collector to detect a shorted condition (should be a very high resistance in either direction). Most transistor failures are either shorted or open elements, and testing in the manner described will isolate the majority of defective devices.

2. The resistance of the emitter-base and collector-base diodes of most transistors will be about the same. Since the diodes appear identical, a transistor could be represented by the symbols shown in Fig. B-2.

   This suggests that the transistor is a symmetrical device, Actually, only certain specially

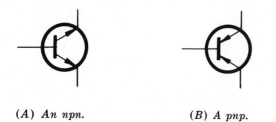

(A) An npn.   (B) A pnp.

Fig. B-2. Base-emitter junction and base-collector junction are actually diodes.

manufactured transistors are completely symmetrical and are referred to as such. The emitter-base and collector-base diodes of most transistors are electrically different, even though their resistance characteristics may be the same. One of the most important differences between the two diodes (from a service viewpoint) is the voltage at which *zener action* will occur.

Zener Action

When a diode is reverse biased as shown in Fig. B-3 and the voltage is gradually increased, a point will be reached where the current increases sharply. Simply stated, the diode conducts backwards and current flows in the direction of the emitter arrow. This does not harm the transistor, providing the circuit resistances are such that the current does not exceed the capability of the diode. The emitter-base diode of most silicon transistors will zener at 7 to 10 volts. The collector-base diode will zener at a much higher voltage.

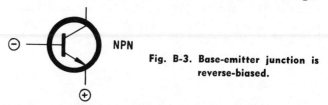

Fig. B-3. Base-emitter junction is reverse-biased.

Most germanium transistors generally conduct backwards (zener) at a considerably higher reverse voltage. However, in germanium transistors when conduction occurs, the current increase is gradual as compared to the silicon type. *Zener diodes*, manufactured for specific operation in this manner, are exclusively silicon.

Probably the most perplexing, but seldom appreciated, problem a service technician encounters is a direct result of the partially symmetrical nature of transistors. A transistor installed backwards (reversed emitter and collector leads) will operate in many circuits at reduced gain. An additionally perplexing effect is that, in most circuits, silicon transistors installed backwards will usually conduct in a zener mode. This can produce voltage readings throughout the associated circuitry that are surprising, to say the least.

Consider how these effects take place. Most transistors operate in circuitry ranging up to approximately 30 volts. Many germanium transistors, because of their higher zener voltage, will operate with collector and emitter leads reversed, but at reduced gain. Silicon transistors, in circuits ranging up to approximately 6 volts, will operate in a similar manner.

If a silicon transistor is connected with the emitter and collector leads reversed in a circuit much in excess of 6 volts, the emitter-base diode will zener. This is understandable when one considers that the emitter-base diode with a low zener potential now occupies the portion of circuitry assigned to the collector-base diode with a higher zener potential. (See Fig. B-4.)

The possibility of improper installation of transistors is being stressed because it occurs more often than is generally appreciated. Transistor basing for similar types varies from one manufacturer to another, and the installation of a replacement transistor backwards, accidentally or because of insufficient information, is very common. This possibility becomes even more common when substitute transistors are used, as is often the case. For example, a service technician is analyzing a direct-coupled audio amplifier. One channel is dead and he quickly determines that the driver transistor is defective. He does not have the original transistor type in stock, but he does have a substitute

transistor recommended in a guide book. The substitute transistor is assumed to have the same basing, but it does not (emitter and collector are

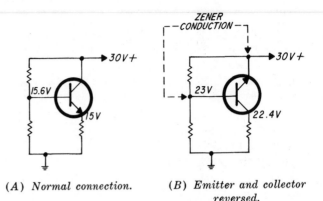

(A) Normal connection.   (B) Emitter and collector reversed.

Fig. B-4. Bias voltages change when the emitter and collector are reversed.

reversed). The circuit configuration is such that, when the replacement transistor is installed, the amplifier operates but the output is reduced and somewhat distorted. The output transistors also run hot after a period of operation. The natural assumption is that since operation has been restored, the problem now lies in some other portion of the circuit.

REVIEW

Questions

Q1. Identify the emitter and collector leads of the transistor in Fig. B-5.

Q2. Is the transistor in Fig. B-5 a germanium or silicon type?

Q3. The circuit illustrated in Fig. B-6 is a level-detection circuit that switches on when the dc voltage present at the input exceeds a certain positive value. The circuit is inoperative. Voltage readings show 9 volts across the emitter-base diode of the transistor. Which component is defective? How and why?

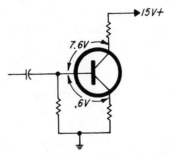

Fig. B-5. Identify the emitter and collector.

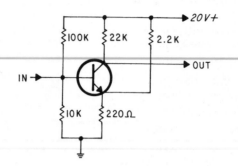

Fig. B-6. A level-detector circuit.

Answers

A1. The emitter lead is toward the top of the page. The base lead is connected through a resistor to ground. If the transistor is an npn, no bias current can flow. If the transistor were a pnp, installed with the emitter lead at the top, the emitter-base diode would be forward biased and a low voltage (tenths of a volt) would be across it. If the transistor were an npn installed normally, the transistor would be cut off (no bias current could flow) and the voltage across the collector-base diode would be about 15 volts. The voltage across the emitter-base diode would be zero since no current would be flowing. The only way the voltage reading shown could occur in this circuit configuration is if an npn transistor were installed with the emitter and collector leads reversed (emitter at the top). The low zener point of the emitter-base diode in the relatively high voltage collector circuit causes the diode to zener and conduct backwards, accounting for the 7.6-volt drop. The 7.6-volt zener voltage also forward biases the collector-base diode, accounting for the .6-volt drop characteristic of a forward-biased silicon diode.

A2. The transistor is of silicon construction. This can be assumed for two reasons:
    (A) The 7- to 10-volt zener voltage is a characteristic of silicon transistors.
    (B) The .6-volt (approximately) forward drop is also characteristic of silicon transistors.
Germanium transistors generally exhibit a .1- to .3-volt forward drop at the current levels produced in most transistor circuits.

A3. The 220-ohm emitter resistor is open. Normally, with no input and all components functioning, the transistor is not forward biased. The resistive divider developing base voltage and the divider developing emitter voltage have the same ratio, producing the same voltage on the base as on the emitter. When the 220-ohm resistor opens, the divider action is lost and the emitter voltage rises to the zener point of the emitter-base diode. The diode conducts backward through the 10k resistor and the 2.2k resistor to B+. The voltage drop produced (9 volts in this case) is the characteristic zener voltage of the emitter-base diode.

C. Transistor Action Can Be Considered as a Small Input Current Controlling a Larger Output Current

This simple concept can be applied as a convenient means of illustrating transistor operation. For the common-emitter configuration, the base-emitter circuit is the input or control circuit. The emitter-collector circuit is the output or controlled circuit. A small input current controls a much larger output current. This is essentially true regardless of the circuit configuration. The signal input and output terminals, and consequently, the input and output impedances, depend on circuit configuration. However, the basic concept of a small current controlling a larger current still applies.

The relation between input current and output current, or more accurately base current to collector current, is called the dc-current transfer ratio (*beta*). The beta also represents the current gain when the transistor is connected as shown in Fig. C-1 (common-emitter configuration). The beta of any particular transistor is a highly variable parameter which is dependent on temperature and operating point. Practical transistor circuits are biased to introduce feedback in such a manner that the effect of beta changes on circuit operation is reduced. Beta is measured at some particular value of collector current. The beta generally becomes less at higher values of collector current.

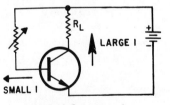

Fig. C-1. Current gain.

Beta may range from 5 to 800, depending on transistor type and at what operating point the measurement is made. A good rule of thumb is to assume a beta of approximately 100 in most low-power (1 mA up to approximately 10 mA collector current) silicon-transistor circuits.

Applications

1. The base current can be used to control the operation of the transistor. This becomes useful because a transistor can be quickly checked in the circuit by increasing or decreasing the base current. The effect of this change is observed in the output (emitter-collector) circuit. If the transistor is operating as a switch and is cut off, attempting to further reduce the base current would prove nothing. In this case, the base current could be increased while observing the transistor output circuit to determine whether or not conduction begins. Conversely, if the transistor is switched completely on (saturated), transistor action can be determined by reducing the base current while observing the transistor output circuit.

2. The voltage appearing across a transistor (emitter-collector) will increase as conduction decreases and decrease as conduction increases. A resistor connected from the base to the collector will increase output conduction because the base current is increased. The resistance value used to effect a substantial change in output conduction would depend on the particular circuit. However, a good rule of thumb is to use a resistor which is one-half the value of the bias resistor in the circuit (the resistor connected from the base to the collector supply). Transistor conduction can be reduced by connecting a resistor between the emitter and the base, reducing the base current. The transistor can be cut off by shorting the emitter to the base. This procedure is permissible in most circuits as a convenient means of cutting off a transistor for test purposes. Care must be exercised in high-power circuits, however, since the sudden cutoff can generate voltage transients which may destroy the device.

3. When the transistor is cut off (not conducting), the full supply voltage will appear across the transistor and zero voltage will appear across the collector load resistor.

4. If there is no base current, the transistor will be cut off and there will be no output current (emitter-collector current).

5. Reducing the value of the base bias resistance in Fig. C-1 will cause the output or emitter-collector current to increase proportionately. Continued reduction of the base bias resistance will further increase emitter-collector current until a point is reached where further reduction of the base resistor value will not cause an increase in the output current. The device, at this point, is said to be *saturated*. The amount of base current required to produce saturation depends on the beta of the transistor and the value

of collector load resistor. The larger the load resistance, the more rapidly saturation occurs; i.e., less base current is required to produce saturation.

6. When a transistor is saturated, the voltage across the transistor will be close to zero. The actual voltage across the device will depend on the amount of collector current, which is determined by the resistance values in the collector circuit.

REVIEW

Questions

Refer to the schematic in Fig. C-2.

Q1. What occurs when R1 is reduced in value? What happens when R1 is open?

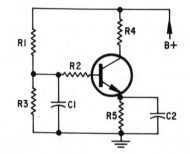

Fig. C-2. An npn transistor circuit.

Q2. C2 is shorted. What happens to the voltage across the transistor (emitter-collector)?

Q3. C1 is shorted. How does this affect transistor operation?

Q4. R3 is open. How does this affect the voltage across the transistor?

Q5. R2 is open. What is the emitter voltage and the collector voltage?

Answers

A1. The input current path consisting of R5, the emitter-base junction R2, and R1 has been made lower in resistance by reducing the value of R1. Therefore, more input current will flow causing a much larger output current through R5, the transistor, and R4. Another way of saying the same thing is: Reducing R1 in value applies a higher positive potential to the base, increasing the forward bias. If R1 is open, no input current flows and the output current is zero.

A2. The voltage across the transistor is reduced. C2 shorts out R5, reducing the resistance of the input circuit and increasing the current flow. Output current also must increase due to the increased input current. Since R4 is fixed in value, the only way output current can increase is for the transistor resistance to decrease. This decrease causes less voltage drop across the transistor and more voltage drop across R4.

A3. The transistor stops conducting. All of the input current now takes the path of least resistance through C1 and R1. No current flows in

14

the input circuit, and therefore, no current flows in the output circuit. The voltage across the transistor rises to B+ potential.

A4. The voltage across the transistor drops. When R3 opens, the divider action of R1 and R3 is lost, causing an increase in voltage at this junction. The base current and, consequently, the emitter-collector current increases. Since R4 and R5 are fixed in value, increased conduction can only take place by lowering the resistance of the transistor. The voltage across the transistor will, therefore, be reduced.

A5. The emitter voltage is zero. The collector voltage will rise to the B+ voltage. When R2 is open, no input current can flow and, as a result, no output current will flow. Since the transistor is an open circuit, the entire power-supply voltage appears across it.

D. Moving the Base Toward the Collector Potential Increases Transistor Conduction and Moving the Base Toward the Emitter Potential Decreases Conduction

The method of developing the base bias, which is common to most circuits, is derived from this concept and is illustrated in Fig. D-1. A divider configuration is generally used to develop the base bias for two reasons.

1. Ease of design: the operating point of the transistor can be quickly approximated by applying certain rules of thumb and simple arithmetic.
2. Impedance matching and stability: a divider configuration allows a wide range of possible resistance values since the ratio determines the voltage developed and, therefore, the base current. (This is provided the resistance is low enough to produce the required base current.)

The resistance values of the base divider greatly influence the stability of the circuit (*S factor*).

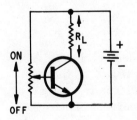

Fig. D-1. Transistor biasing circuit.

Generally, the lower the resistance of the base divider, the more stable the circuit becomes. Also, the current gain of the stage is correspondingly reduced because of the lower impedance of the input circuit.

Applications

1. Practical biasing circuits usually consist of a resistor voltage divider. Increasing the value of the base-to-emitter resistor causes increased transistor conduction. Decreasing the value of the base-to-collector resistor also causes increased conduction.
2. When the base is at emitter potential (zero voltage between base and emitter), the transistor is cut off (no conduction).
3. Assume that the control wiper in Fig. D-1 is positioned at the OFF end of the control. Moving the wiper slightly up toward ON will supply a small positive potential to the base. However, significant emitter-collector conduction will not take place until about 0.6 volt (in the case of a silicon transistor) is applied to the base. The forward-bias junction potential of the base-emitter diode must be overcome before conduction can occur.

REVIEW

Questions

Refer to Fig. D-1.

Q1. What will be the effect if a very high resistance is used as the variable base control?

Q2. What will be the effect if a very low resistance is used as the variable base control?

Q3. Do you think the resistance value selected for this control would have any effect on the input signal?

Answers

A1. The loading effect of the base-emitter diode would require that the sliding contact be moved toward the ON position to produce the same base voltage and input current that would be obtained from a smaller resistance. A high resistance will allow greater gain at the expense of stability.

A2. The sliding contact would have to move toward the OFF position to achieve the same conduction that would be produced by a larger resistance. The low value would load the power supply by a greater amount and result in wasted power. This would be important if battery power were being used. The lower resistance will also increase the stability of the stage at the expense of gain.

A3. Yes, the lower the resistance of the control, the greater the loading effect on the input signal. The signal sees an impedance that is the total parallel impedance of the emitter-base junction, the base-emitter resistor, and the base-collector resistor. This is true because the signal sees ground and B+ as at practically the same ac potential due to the low impedance of the filter capacitors.

E. The Output (Emitter-Collector) Circuit of a Transistor Can Be Considered a Variable Resistor

As far as practical servicing is concerned, the value of this resistor can range from almost a dead short to an open circuit (Fig. E-1). The resistance value is determined by the amount of bias or quantity of current flowing in the base-emitter circuit.

Applications

1. When no base-emitter current is flowing, the value of this variable resistor is, for all practical purposes, infinite or open. No current can flow and the full value of the supply voltage will appear across the emitter-collector circuit. The effect would be the same as if any of the transistor elements were open internally.

2. When the base-emitter bias current is increased to the point where no further change occurs in output current, the transistor is said to be *saturated*. The point at which this occurs depends on the beta of the transistor and the value of the collector load resistor. Saturation merely means that the transistor output (emitter-collector) resistance has been reduced to its lowest

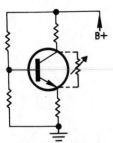

Fig. E-1. A transistor can be considered a variable resistor.

possible value. A saturated transistor can be considered the same as a turned-on switch. The resistance value is very low and usually less than 1 volt will appear across the device. The voltage will depend on the magnitude of the emitter-collector current, which is also dependent on the external resistance values.

3. Different transistor types and different collector load-resistor values will require more or less input (base) current to reduce the emitter-collector resistance to its lowest value. A typical value of base current to produce saturation in an audio-driver transistor could be about 1 mA. A power transistor may require up to 100 mA of base current to produce saturation. Saturation, to be meaningful, must be specified at some value of collector load resistance. The base currents (1 mA and 100 mA) specified above would be required only if the respective collector currents were very high and the load resistance very low. These collector currents may be on the order of 100 mA for the audio-driver transistor and several amperes for the output transistor. At lower values of collector current and higher values of load resistance, saturation is achieved at proportionately lower values of base current.

Generally, a transistor dissipates the least amount of power when operated as a switch (either open or saturated). When open, no current flows; when saturated, the emitter-collector resistance is at the lowest possible value and little power is dissipated. This is assuming, of course, that the external circuit resistances are such that excessive current does not flow. A good example of this is the horizontal sweep circuitry of most solid-state tv receivers. The horizontal-driver and output transistors are switched into saturation and cutoff. Operating in this mode, the time between saturation and cutoff is when the greatest power dissipation within the transistor occurs. If this time is too long, the transistor overheats and fails.

4. When operating as a class-A amplifier, the transistor is operated somewhere between cutoff and saturation and a portion of the power-supply voltage will appear across it. A transistor, biased so that one-half of the supply voltage appears across the collector load resistor and the other half across the transistor and the emitter resistor, would be operating exactly in the middle of its class-A range. At this time, the resistance from emitter to collector would be equal to the collector load resistor minus the emitter resistor. An ac signal driving the base of the transistor can now increase or decrease this resistance an equal amount in either direction.

REVIEW

Questions

Examine the schematics in Fig. E-2.

Q1. The transistor in Fig. E-2A is saturated. What is the approximate voltage across the 1k resistor? What is the approximate collector current?

Q2. What is the emitter-to-collector voltage of the transistor in Fig. E-2D?

Q3. Is the transistor in Fig. E-2B saturated?

Q4. The emitter-to-collector voltage across the transistor in Fig. E-2C is 5.5 volts. What is the class of operation? What will the voltage be if the 22k resistor opens?

Q5. If C1 in Fig. E-2D shorts, what is the emitter-to-collector voltage across the transistor?

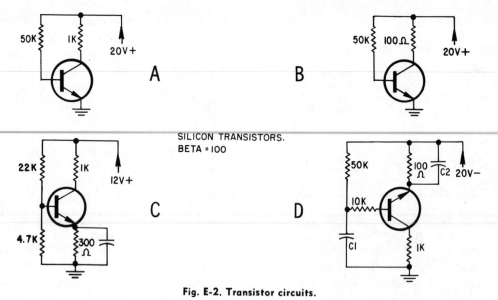

Fig. E-2. Transistor circuits.

Q6. The collector voltage relative to ground in Fig. E-2C is 7 volts. What is the emitter-to-collector resistance of the transistor?

Answers

A1. The voltage across the resistor will be approximately 20 volts. The transistor, when saturated, acts as a closed switch of negligible resistance compared to the rest of the circuit. The current flowing in the circuit is limited primarily by the resistor value. The collector current would be about 20 mA.

$$I = \frac{E}{R}$$
$$= \frac{20}{1000}$$
$$= .02 \text{ ampere}$$
$$= 20 \text{ mA}$$

A2. 20 volts will appear across the transistor. Notice that the positive end of the power supply is grounded. Electron flow from negative to positive through the emitter-base circuit is impossible because C1 blocks the flow of dc. The transistor is cut off, presenting an open circuit across the power supply.

A3. No. We can determine this by making several rule-of-thumb assumptions. The emitter-base resistance is small compared to the 50k resistor and so it can be ignored for all practical purposes. Base current is then:

$$I = \frac{E}{R}$$
$$= \frac{20}{50,000}$$
$$= .0004 \text{ ampere.}$$

Assuming a beta of 100, which is not unreasonable, the output current would be .004 × 100 or .04 ampere. The voltage across the load resistor would be:

18

$$E = IR$$
$$= .04 \times 100$$
$$= 4 \text{ volts.}$$

Obviously, the transistor is not saturated. Simple calculation would show that a beta of at least 500 or a 10k base-bias resistor would be required to achieve saturation in this circuit. With these values, there would be a 20-volt drop across the load resistor. This would leave zero voltage across the transistor, indicating saturation.

A4. The transistor is operating class-A. A portion of the power-supply voltage appears across the transistor. The transistor is neither saturated nor cut off. Opening the 22k resistor will remove the bias current and the transistor will be cut off. The full supply voltage, or 12 volts, appears across the transistor.

A5. Less than 1 volt will appear across the transistor. The transistor is saturated. The total resistance of the input circuit can be approximated by simple rule-of-thumb calculations. The input resistance of the common-emitter configuration is approximately the resistance of the emitter-base diode plus the emitter resistor times beta, or:

$$R_{eb} + R_e \times \text{beta} = \text{input resistance.}$$

The small emitter-to-base resistance can be represented by subtracting the 0.6-volt diode drop from the supply voltage. Therefore, the emitter-base circuit resistance becomes:

$$\text{beta} \times 100 \text{ or } 10,000 \text{ ohms.}$$

Adding this to the 10k resistor in the base circuit, we have a total input resistance of 20k ohms. The base circuit is, therefore:

$$I = \frac{E}{R}$$
$$= \frac{20 - 0.6}{20,000}$$
$$= \frac{19.4}{20,000}$$
$$= .001 \text{ ampere approximately}$$
$$= 1 \text{ mA.}$$

Assuming a beta of 100, the output (emitter-collector) current would be .1 ampere (100 mA). The total output resistances are such that only 18 mA of current can flow.

$$I = \frac{E}{R}$$
$$= \frac{20}{1100}$$
$$= .018 \text{ ampere.}$$

Since the input circuit values are such that up to 100 mA of output current can be produced, we can safely assume that the beta and bias values are more than adequate to create a saturated condition. The beta or the bias current would have to be diminished by a factor of at least six times before a class-A operating point would be established.

A6. 1100 ohms. From Kirchhoff's law, we can determine that 5.5 volts appear across the transistor. The 5-volt drop across the 1k collector resistor gives a collector current of 5 mA. Assuming that the emitter and collector currents are approximately the same, we have by Ohm's law:

$$R = \frac{E}{I}$$
$$= \frac{5.5}{.005}$$
$$= 1100 \text{ ohms.}$$

F. The Input (Base-Emitter) Circuit of a Transistor Is a Diode

Notice that this statement does not say the input circuit of a transistor may be represented as a diode or that the input circuit acts like a diode. The input circuit *is* a diode. This may seem an insignificant point, but it is stressed because many technicians and much information written about transistor operation seem to overlook this fact.

The electrical characteristics of a diode primarily depend on its material makeup, whether silicon or germanium. Most germanium diodes begin significant conduction in the forward direction when a potential of 0.1 volt or greater is applied. Most silicon diodes begin significant conduction at 0.5 volt. Application of voltages higher than these results in a rapid reduction of the internal resistance of the diode and a sharp rise in current. Because of these characteristics, the voltage drop as measured across the diode remains fairly constant. All diodes are essentially voltage regulators or constant-voltage devices when biased in the forward direction. Silicon diodes are better in this respect than germanium diodes because of the sharper knee, as illustrated in Fig. F-1.

The application of reverse bias causes no significant conduction until the zener voltage is reached. The diode then conducts suddenly and again becomes a voltage regulating device—a zener diode. The zener voltage of most silicon emitter-base diodes is 7 to 10 volts. Notice in Fig. F-1 that a germanium diode does not have a well-defined zener region.

Applications

1. A forward-biased transistor will maintain an almost constant voltage across the emitter-base diode over its entire operating range. The im-

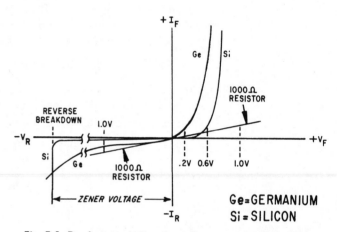

Fig. F-1. Dc characteristics of a resistor, a germanium diode, and a silicon diode.

portance of this voltage is often stressed, indicating that it represents the bias and establishes the operating point of the transistor. While true, this is also a very misleading and, in many respects, a useless concept. Many technicians religiously measure this voltage until they realize that the operating point cannot be determined from the measurement. The amount of emitter-to-base voltage change from near cutoff to saturation is insignificant and can provide no meaningful information to the technician. Proper emitter-to-base voltage indicates only that the diode is forward biased, not the degree of forward bias.

2. Most silicon transistors maintain a base-to-emitter voltage of about 0.6 volt.

3. Most germanium transistors maintain a base-to-emitter voltage of from 0.1 to 0.3 volt. Germanium transistors exhibit a wider diode voltage variation because the conduction curve does

not have as sharp a knee as the silicon diode. These curves are illustrated in Fig. F-1.

4. Many useful rule-of-thumb approximations can be devised concerning circuit operation by applying this relatively constant diode voltage drop in various calculations. For example, if the base voltage in a particular circuit is known, the emitter voltage can also be approximated and vice versa. This principle will be further illustrated in the questions and answers that follow.

5. The voltage drop across a diode varies slightly with temperature. For this reason, diodes are often used in transistor biasing circuits to provide stabilization of the operating point.

6. Whatever appears on the base of a forward-biased transistor also appears on the emitter and vice versa. Assuming that the signal does not overcome the dc bias (class-A operation), the signal appearing on the emitter will be very much the same as the signal on the base. This is particularly true if the emitter resistor is not bypassed with a capacitor. If the emitter is bypassed, the signal on the emitter appears slightly lower in amplitude than the signal on the base.

7. A transistor that does not exhibit this diode voltage across the base-emitter junction is not forward biased and should not be conducting. An exception to this is class-C operation, where signal rectification in the base-emitter circuit can cause a reverse voltage to appear across the diode.

REVIEW

Questions

Refer to Fig. F-2.

Q1. What is the voltage relative to ground at point A in Fig. F-2A?

Q2. What is the approximate collector current of the transistor in Fig. F-2D?

Q3. What is the voltage relative to ground at point A in Fig. F-2B?

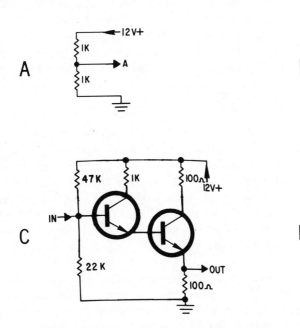
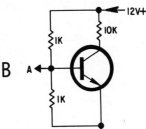
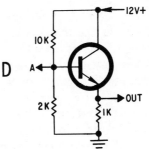

Fig. F-2. Simplified transistor circuits.

Q4. What is the difference in amplitude between the input and output signals in Fig. F-2C? What would the difference be if the emitter resistor were 10 ohms?

Q5. If the 2k resistor in Fig. F-2D opens, what voltage can we expect at the output?

Q6. The emitter-to-base voltage in Fig. F-2D reads 2 volts. What is the state of conduction of the transistor? What do you think the problem is?

Q7. What is the voltage at point A in Fig. F-2B if the lower 1k resistor opens?

Q8. Are the transistors in Fig. F-2C forward biased? Why?

Q9. What is the minimum voltage which must be present at the input in Fig. F-2C to forward bias both transistors?

Q10. What is the voltage across the output transistor (emitter to collector) in Fig. F-2C?

Q11. How much current would flow in the output circuit in Fig. F-2B if the transistor were shorted from emitter to collector? How much is flowing in it as shown?

Answers

A1. 6 volts. The divider resistors are equal in value and, therefore, half the supply voltage appears at point A.

A2. About 1.4 mA. The base divider resistors produce about 2 volts at the base. The voltage produced by any resistive divider network can be easily determined by a simple ratio relationship. The ratio of either resistor over the total resistance is equal to the ratio of the voltage across the selected resistor over the total supply voltage. In this case, we are interested in the base voltage with respect to ground, or the voltage across the 2k resistor. The ratio would therefore be:

$$\frac{2000}{12,000} = \frac{X}{12}$$

The X, or unknown voltage, can now be determined by cross multiplying, and we derive 12,000X equals 24,000 or X equals 2 volts.

The base voltage will actually be slightly less than 2 volts because of the loading effect of the input circuit. The loading effect in most practical circuits is negligible, and for all practical purposes we can say the base voltage is 2 volts. (In this case the loading is about 100,000 ohms or beta times the 1k resistor.)

The transistor is forward biased. This means that there is a 0.6-volt drop across the base-emitter diode. The emitter voltage can be determined by subtracting the 0.6-volt drop across the base-emitter diode from the base voltage. The emitter voltage is 1.4 volts. We can now derive emitter current by applying Ohm's law.

22

$$I = \frac{E}{R}$$

$$= \frac{1.4}{1000}$$

$$= .0014 \text{ ampere}$$

$$= 1.4 \text{ mA.}$$

The collector current in most practical circuits is almost the same as the emitter current. (Collector current equals the emitter current minus the small base current.) Therefore, we can safely say that the collector current in this circuit is 1.4 mA.

A3. The emitter is connected directly to ground. The diode is forward biased and must exhibit the characteristic 0.6-volt silicon diode drop. Therefore, the base voltage at point A must be 0.6 volt. This example may seem to negate the statement made in A2 that the loading effect of the input circuit is negligible. Remember that this assumed a practical circuit with negligible loading of the base divider resistors. The example shown in Fig. F-2B loads the base circuit considerably and is not a practical circuit by any means. The circuit values are, in fact, such that there will be more input current than output current.

A4. The signal at the output will be slightly less in amplitude than the input. Calculation will show that both transistors are forward biased. The amplitude of the input signal will be greater than the output signal by the voltage developed across the relatively low resistance of the forward-biased diodes. The difference in signal amplitudes would be about the same if the emitter resistor were 10 ohms. The input impedance of such a circuit can be very high and is usually about the value of the two bias resistors in parallel.

A5. About 10 volts. When the 2k resistor opens, the divider action is lost and the current that will flow in the input circuit is proportional to the total input resistance. Ignoring the small base-emitter diode resistance, this will be:

$$\text{beta} \times 1k + 10k = 110k \text{ ohms.}$$

The total current will be:

$$I = \frac{E}{R}$$

$$= \frac{12}{110k}$$

$$= .0001 \text{ ampere.}$$

The drop across the 10k resistor under these conditions will be 1 volt.

$$E = IR$$

$$= .0001 \times 10,000$$

$$= 1 \text{ volt}$$

Therefore, the voltage applied to the base will be 11 volts. Deducting the silicon diode drop (0.6 volt) will result in 10.4 volts at the emitter.

A6. The transistor is cut off. The voltage drop across a normal forward-biased silicon diode cannot be 2 volts. The only problem possible is that the emitter-base diode is open.

A7. The voltage at point A will be 0.6 volt. The lower 1k resistor, in parallel with the emitter-base diode, normally has little effect on the low forward resistance of the diode anyway.

A8. Yes. The voltage applied to the base of the first transistor can be determined from the ratio relationship previously discussed.

$$\frac{X}{12} = \frac{22}{69}$$
$$69X = 264$$
$$X = 3.8 \text{ volts}.$$

The loading effect of the input circuit is negligible since its resistance is beta times beta times 100 or 1 megohm. The base voltage of the second transistor would be about 3.8 − 0.6 or 3.2 volts. Both transistors are therefore forward biased.

A9. The voltage must be in excess of 1.2 volts (0.6 volt is required for each diode).

A10. About 6.8 volts. The base divider produces 3.8 volts. (See A8.) The base voltage of the second transistor is 3.2 volts. The emitter voltage must then be 2.6 volts (3.2 − 0.6). The emitter current is, therefore:

$$I = \frac{E}{R}$$
$$= \frac{2.6}{100}$$
$$= .026 \text{ ampere}$$
$$= 26 \text{ mA}.$$

The collector current, being approximately the same as the emitter current, will produce a voltage of 2.6 volts across the 100-ohm collector load resistor. The 2.6 volts across the collector load resistor and 2.6 volts across the emitter resistor leaves 6.8 volts (12 − 5.2) across the transistor.

A11. The current will be .0012 ampere (1.2 mA). The current is limited only by the load resistor.

$$I = \frac{E}{R}$$
$$= \frac{12}{10,000}$$
$$= .0012 \text{ ampere}.$$

As shown, approximately the same current of .0012 ampere is flowing since the transistor is saturated.

G. The Base and Emitter of a Forward-Biased Transistor Will Always Follow Each Other, Both Regarding DC Potential and AC Signal

The base-emitter circuit of a transistor is a diode, usually operated in the forward-biased mode. When the diode is forward biased, its impedance can be represented by an interesting and important equation known as *"Shockley's Relation."* (Shockley is one of the inventors of the transistor.) This equation states that the base-emitter diode impedance is equal to 26 ohms divided by the emitter current expressed in milliamperes. Surprisingly, this holds true for both silicon and germanium transistors. In actual practice, this formula usually works out to be closer to 30 ohms divided by the emitter current in milliamperes and will be represented as such in this book.

Normally, this value is considered only as an ac impedance. However, at any given instant, ac is a small change in dc level. Therefore, Shockley's Relation may also be considered a dc resistance to small changes in dc voltage for a forward-biased diode.

Most circuits encountered in home entertainment products are designed to draw at least 1 mA of emitter current. The diode impedance then becomes 30 ohms or less. Compared to the other resistances in the circuit, this is usually quite low.

Applications

1. When forward biased, the base and emitter voltages will be very close, regardless of the external base or emitter circuit values (about 0.6 volt for silicon and 0.1 to 0.3 volt for germanium). If this relationship does not exist, the circuit design is such that the transistor is not forward biased or the diode junction is defective.
2. When a transistor is forward biased, a change in base voltage produces an almost identical change in emitter voltage.
3. A change in emitter voltage will usually cause the base voltage to change. The degree of tracking would depend on the resistance of the base bias supply compared to the resistance of the supply causing the change in emitter voltage.
4. When a transistor is forward biased, the signal appearing on the base and emitter will be nearly equal. This is particularly true if the emitter resistor is not bypassed with a capacitor and is much larger than the emitter-base diode impedance. When the emitter is connected to ground or is ac grounded (bypassed), the signal on the base will be proportional to emitter current. The base-emitter diode impedance will be about 30 ohms divided by the emitter current in milliamperes. This impedance multiplied by beta will be the impedance across which the signal will appear (common-emitter configuration). A practical illustration of this principle is the electronic-filter circuit. A power transistor is forward biased to operate in a linear (class-A) manner. The ac ripple is removed from the base circuit by a relatively low value capacitor, and since the emitter follows the base, ripple is also removed from the emitter circuit. The small amount of current in the base circuit is easily filtered, resulting in a large amount of filtered current in the emitter circuit (emitter-collector circuit). The effectiveness of such a circuit is approximately equal to beta multiplied by the value of the base capacitor in microfarads.
5. When the output signal is taken off at the emitter, the circuit is called an *emitter follower*. Sometimes this is also called a common-collector circuit, which may or may not be the case depending on circuit configuration. An emitter follower never exhibits voltage gain since the signal on the emitter is always slightly less than the signal on the base.

REVIEW

Questions

Examine the schematics in Fig. G-1.

Q1. In which circuit shown in Fig. G-1 will the emitter and base signals be more nearly equal?

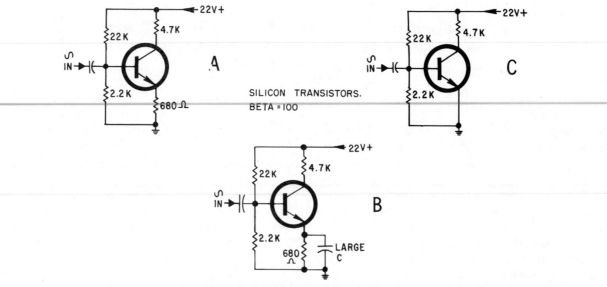

Fig. G-1. Typical transistor circuits.

Q2. Assume that the emitter bypass capacitor in Fig. G-1B is large enough to act as an ac short at the particular frequency being amplified. What is the approximate ac resistance (impedance) of the input circuit?

Q3. How does the ac input impedance of the circuit in Fig. G-1B compare to the circuit in Fig. G-1C.

Q4. Assume that the input capacitors have a high impedance at the particular frequency being amplified (say on the order of 10k ohms). An identical signal is applied to the three circuits. Will the amplitude of this signal be the same on each base?

Q5. How much voltage gain can be expected from an emitter-follower circuit?

Q6. If the emitter and base voltages differ more than one volt, what two conclusions could be reached?

Answers

A1. Fig. G-1A. The impedance of the emitter-base diode is low in comparison to the resistors in the circuit. The dc operation of the circuit and the value of the emitter-base resistance can be easily approximated as follows:

The base voltage, ignoring the slight loading effect of the base-emitter circuit, is the divider voltage across the 2.2k resistor.

$$\frac{X}{22} = \frac{2,200}{24,200}$$
$$X = \frac{2,200 \times 22}{24,200}$$
$$= 2 \text{ volts.}$$

The emitter voltage will be 0.6 volt less than the base voltage, or 1.4 volts. The emitter current will be:

26

$$I = \frac{E}{R}$$

$$= \frac{1.4}{680}$$

$$= .002 \text{ ampere}$$

$$= 2 \text{ mA.}$$

The emitter-base diode resistance is then 30 ohms divided by the emitter current in milliamperes or:

$$\frac{30}{2} = 15 \text{ ohms (Shockley's Relation).}$$

The voltage across the load resistor is:

$$E = IR$$

$$= .002 \times 4700$$

$$= 9.4 \text{ volts.}$$

The input resistance, or impedance in the case of ac, would be about beta times the emitter resistor plus the emitter-base diode resistance or:

$$100 \times (680 + 15) = 69,500 \text{ ohms.}$$

The base-bias resistors have, in this case, been ignored for the sake of illustration. In the circuits shown, the bias resistors are of much lower value than 69,500 ohms and, as such, primarily determine the actual input impedance. Circuit values of this sort are typical of low gain, high stability amplifiers.

Notice that the value of the emitter-base impedance (15 ohms) in relation to the emitter resistor (680 ohms) is very low. The amount of signal voltage drop across the emitter-base junction will, for this reason, be negligible.

A2. 1500 ohms (the bias resistors have again been ignored to illustrate the method of determining the impedance of the emitter-base circuit). The dc conditions of the circuit in Fig. G-1B are identical to the circuit in G-1A. The dc conditions described in A1 dictate an emitter-base impedance of 15 ohms. The bypassed emitter resistor has no impedance to the ac signal. The ac input resistance is, therefore, 15 × beta, or 1500 ohms.

A3. The impedance of the circuit in Fig. G-1C is lower. At first glance, it may seem that the ac conditions are the same in circuits B and C. This is not, in fact, the case because the change in dc operating point also causes a change in both ac and dc input resistance. The emitter-base diode resistance will be much less than the 2.2k resistor which can, for practical purposes, be considered out of the circuit. The base circuit current will be approximately 1 mA.

$$I = \frac{E}{R}$$

$$= \frac{21.4}{22,000}$$

$$= .001 \text{ ampere}$$

$$= 1 \text{ mA.}$$

The maximum collector current could be:

$$100 \text{ (beta)} \times .001 = 100 \text{ mA.}$$

Since the collector load resistor will not allow nearly this much current flow, we can safely say the transistor is saturated. The maximum emitter-collector current will be:

$$I = \frac{E}{R}$$
$$= \frac{22}{4700}$$
$$= .004 \text{ ampere}$$
$$= 4 \text{ mA.}$$

The emitter-base impedance will be somewhat less than:

$\frac{30}{4}$ or 7.5 ohms (Shockley's Relation).

The input impedance will be 7.5 × beta, or 750 ohms. This is half the impedance of the circuit in Fig. G-1B.

A4. No. The base-emitter circuit will load this signal. The signal will be of greatest amplitude on the base of the circuit in Fig. G-1A and of lowest amplitude in the circuit in Fig. G-1C.

A5. None. Some voltage loss will be encountered due to the impedance of the emitter-base diode. Power amplification is realized in such a circuit due to current gain.

A6. The diode is either reverse biased or open.

H. The Voltage Appearing Across the Transistor Is the Most Significant Test Measurement That Can Be Made

Voltage measurements between the three transistor elements indicate the dc operation of the device. The most informative of these measurements is the voltage from emitter to collector. When a stage is suspected of being faulty, the first tests to be made concern the dc operation. The only measurement that conveys this information is the emitter-to-collector voltage appearing across the transistor. This voltage quickly indicates whether the transistor is cut off, saturated, or operating somewhere between (class-A).

For the sake of illustration, consider what voltage readings at or between the other elements indicate. Why not just measure the collector voltage? This is fine except the collector may be grounded or connected directly to B+, and thus indicate nothing significant. The same situation exists if only emitter voltage is measured.

What does measurement of base voltage tell us? The base voltage will be the emitter voltage minus (pnp) or plus (npn) the voltage drop across the diode, assuming the transistor is forward biased and the base-emitter junction is not open or shorted. The base voltage indicates the bias condition of the transistor and is probably the second most significant measurement that can be made. Transistor output current is proportional to input current (beta ratio). We can, therefore, say that transistor conduction is controlled by the voltage on the base and the value of the emitter resistor (more accurately, the emitter resistor plus the base-emitter diode resistance). Measurement of the emitter-to-collector voltage of the transistor indicates the operating mode of the transistor and measurement of the base voltage generally tells you why it is conducting in this manner.

Many technicians stress the significance of the voltage appearing between the base and emitter terminals. This voltage indicates only the state of bias, not the degree. This voltage is the drop across a forward-conducting diode, and it remains virtually unchanged from near transistor cutoff to saturation. This measurement can be misleading, even concerning the state of bias, since a transistor which is driven class-C (as a switch) appears to be reverse biased. The implication here is not that this voltage measurement is useless, but rather that the measurement can be made to determine an open or shorted diode. This measurement will also quickly indicate a diode in zener conduction. A reading of 7 to 10 volts across this diode almost always indicates zener conduction for a silicon transistor.

Applications

1. The more a transistor is forward biased, the lower its resistance will be from emitter to collector. This will give a lower emitter-to-collector voltage across the transistor.

2. The full supply voltage will appear across the transistor (emitter-to-collector) if it is not conducting. This is assuming that the cutoff state is caused by the bias supply, an open element within the transistor, or a shorted emitter-base diode. The supply voltage will not appear across the transistor if the external circuit is open (emitter or collector resistor). If the collector resistor is open, the voltage across the transistor would be zero. An open emitter resistor can be confusing since a portion of the power-supply voltage will appear across the transistor, suggesting proper class-A operation. The magnitude of this voltage depends on several things:

   (a) The reverse leakage of the emitter-base diode.
   (b) The impedance of the meter being used.
   (c) The supply voltage. (The higher the supply voltage, the closer the emitter-base diode approaches the zener point.)
   (d) The amount of signal present. When measured with a high-impedance meter, signal detection by the base-emitter diode also contributes to the dc voltage reading across the transistor.

3. When biased fully on (saturated), about 1 volt or less will appear across the transistor. (This will also occur if the collector load resistor is open.) The external resistance values in most low-powered circuits are such that, when saturated, the voltage across the transistor is zero for all practical purposes. Saturation essentially means that the resistance value of the collector-emitter circuit has been reduced to a very low value in comparison to the collector load resistor. Many technicians have an inaccurate impression of just what saturation implies. When questioned on this subject the answer most often heard is: "The transistor is conducting like mad." Conduction refers to current flow and the transistor may indeed be "conducting like mad" or it may be conducting very little. The amount of conduction depends on the resistance values in the external collector-emitter circuit.

4. The full supply voltage should appear across the transistor if the forward bias is removed. This quickly distinguishes between a saturated and shorted transistor. A convenient means of doing this is to short the base to the emitter. This is permissible in most circuits. However, caution must be exercised in high-power circuits, especially those containing inductive components. The sudden transistor cutoff may generate voltage transients that can damage solid-state devices.

5. A portion of the supply voltage will appear across the transistor when it is operating class-A. The transistor is operating in the middle of its class-A range if half the supply voltage appears across the collector load resistor. Maximum signal swing can be achieved in this mode. When the collector load is inductive, most of the supply voltage usually appears across the transistor due to the low dc resistance of the load. The operating point, however, may still be a class-A as a result of the ac impedance of the load.

6. All voltage and resistance measurements have some meaning depending on the situation. The emitter-to-collector voltage across the transistor is recommended simply because it supplies the greatest amount of information in the greatest diversity of circuit configurations and possible malfunctions. The two exceptions which require additional voltage or resistance measurements are an open collector resistor or an open emitter resistor.

REVIEW

Questions

Examine the illustrations in Fig. H-1.

Q1. Assume E = 12 volts in Fig. H-1A. List all possible reasons why this could be so.

Q2. Assume E = 0.06 volt in Fig. H-1A. List all possible reasons why this could be so.

Q3. What do you think E (Fig. H-1A) would be if everything were operating properly?

Q4. Assume E = 0.6 volt in Fig. H-1A. Short the base to the emitter. E remains at 0.6 volt. What do you conclude from this?

Q5. What is the voltage across the second transistor in Fig. H-1B?

Q6. What is the voltage across the second transistor in Fig. H-1B if the emitter bypass capacitor is shorted?

Q7. What is the voltage across the second transistor in Fig. H-1B if the diode is reversed? What is the emitter voltage?

Answers

A1. (a) The emitter-base diode is shorted. The transistor stops conducting and the voltage rises to B+ potential.

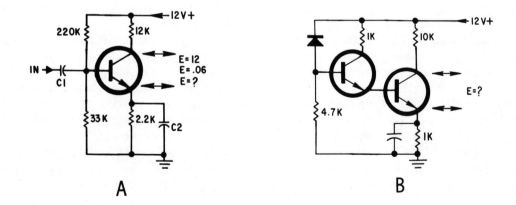

Fig. H-1. Analysis of the emitter-to-collector voltage of transistor circuits.

(b) E would equal 12 volts if any of the transistor elements were open internally. Any open element would cause the transistor to stop conducting and the voltage would rise to the B+ supply potential.

(c) The 220k resistor is open. This would remove forward bias and the transistor would stop conducting.

(d) A shorted 33k resistor. This would also remove forward bias and the transistor would stop conducting.

A2. (a) The transistor is shorted from emitter to collector. The 0.06 volt is due to resistive voltage drop (may not be a dead short).

(b) The 33k resistor is open. Calculation will show that sufficient bias current will not flow to saturate the transistor. The voltage across the transistor will, as a result, approach zero.

(c) The 220k resistor is shorted. Resistors normally do not short, but a solder bridge across two foil conductors on the printed-circuit board could easily cause the equivalent. Again, calculation will show that the transistor is saturated.

(d) C2 is shorted. With C2 shorted, the 33k resistor is essentially out of the circuit. The input current will approximate:

$$I = \frac{11.4}{220k} = .00005 \text{ ampere.}$$

The potential output current is this value times beta, or about .005 ampere. The maximum output current possible with a 12k resistor is .001 ampere. We can thus safely assume the transistor is saturated and the voltage across it will approach zero.

(e) C1 could be shorted. In most circuits, C1 would probably be connected to the collector of the preceding stage. The additional positive voltage available to the base circuit if C1 shorts could saturate the transistor.

(f) The collector load resistor is open. A high-impedance meter may read some slight voltage under this condition due to signal rectification.

NOTE: Saturation can be easily determined by simple calculation as illustrated. In most practical circuits, however, it can generally be assumed that saturation will occur if the following conditions prevail.

(1) The emitter-base bias divider resistor or circuit opens.
(2) The collector-base bias divider resistor or circuit shorts.
(3) The emitter resistor or bypass capacitor shorts.

A3. About 5 volts. The bias divider is producing about 1.6 volts.

$$\frac{X}{12} = \frac{33k}{253k}$$
$$X = 1.57 \text{ volts.}$$

Ignoring base current, the emitter voltage will be 0.6 volt less, or 1.0 volt. Emitter current is:

$$I = \frac{E}{R}$$
$$= \frac{1.0}{2.2k}$$
$$= .00045 \text{ ampere.}$$

Voltage across the collector load will be approximately .00045 × 12k, or 5.5 volts (again, ignoring base current). The 5.5 volts across the load resistor and the 1.0 volt across the emitter resistor leave 4.5 volts across the transistor.

A4. The transistor is shorted from emitter to collector. Normally, shorting the base to emitter will cause the transistor to stop conducting, and the voltage across the transistor will rise to the value of the supply voltage. The fact that this did not occur indicates the transistor must be shorted from emitter to collector.

A5. 12 volts will appear across the transistor. The diode is connected so that no base bias current can flow. Both transistors are cut off as a result and the voltage across the transistor rises to the B+ potential.

A6. Again, 12 volts will appear across the transistor. No bias current is flowing because of the reversed diode and shorting the emitter resistor can have no effect.

A7. The voltage across the transistor will approach zero. Calculations will show the transistor is saturated. The emitter voltage must equal the supply voltage minus the drop across the three forward-biased diodes, or 10.2 volts.

$$12 - 1.8 = 10.2.$$

I. When a Transistor Is Driven Into Cutoff or Saturation by a Signal, It May Appear to Be Reverse Biased

This may seem to be a very obvious effect, but apparently this action is not clear to many technicians. This is not surprising since the actual mechanism causing this effect is more complex than generally realized. A further complication is that, although the dc bias shifts in the same direction when driven into cutoff or saturation, the reasons for these shifts are not identical. We will consider the effects caused by driving a transistor into cutoff:

When a transistor is operating class-A, a certain amount of dc forward bias is applied. The signal level at the base must, of necessity, have a lower peak voltage than the dc bias level or class-A operation will no longer occur. If the peak signal is equal to the dc bias voltage, it will cut the transistor off when swinging in the reverse-bias direction. Under these conditions, the measured dc bias will change proportionally to signal strength and the relative impedances of the signal source and the transistor input circuit. The questions following

this section more graphically illustrate this fact. The voltage across the emitter-base diode will begin to diminish as soon as the signal is of sufficient amplitude to drive the transistor into cutoff. This voltage will continue to diminish as the signal is increased and may even pass through zero and reverse polarity. The degree of bias change is very much dependent on the relative impedances of the signal source and the transistor input circuit, as previously stated. A high signal-source impedance (or base resistance) and low input-circuit impedance (or emitter resistance) will produce a large change in dc bias. The opposite situation will produce almost no change. Circuits which make use of this developed voltage (signal biased) are, for this reason, usually operated with no emitter resistor.

The change in dc bias appears as a result of signal rectification. The base-emitter junction of a transistor is a diode and behaves as any other diode when ac is applied. An example of this action is the ordinary silicon rectifier used in most power supplies. When operating, a positive voltage is developed on the cathode of the diode. The voltage across the diode is limited by circuit impedances and zener action. The emitter-base diodes of most silicon transistors zener at about 7 to 10 volts. As a result, the average measured bias change for a silicon transistor due to signal rectification seldom exceeds a few volts.

When a transistor is driven into saturation, the dc bias voltage also changes. In most circuits, this change is more dramatic than the change due to signal rectification by the emitter-base diode. When saturated, the resistance from emitter to collector is very small. The collector and emitter voltages become essentially the same and the collector-base diode becomes forward biased. The effect is as if two diodes were connected in parallel as shown in Fig. I-1.

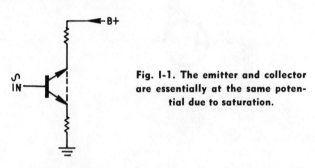

Fig. I-1. The emitter and collector are essentially at the same potential due to saturation.

The change in bias voltage due to signal rectification by the emitter-base circuit may be insignificant, dependent on circuit impedances. However, the shift in bias due to saturation is readily apparent. The reason for this is again a matter of relative impedances. When operating class-A, the input impedance and dc resistance of the emitter-base circuit is approximately beta times the value of the emitter resistor. When saturation occurs, this input impedance is lost due to the shunting action of the forward-biased collector-base diode. This low impedance now loads the signal source during conduction cycles and produces a reverse dc

bias voltage. This is about the same effect as when the transistor is operated without an emitter resistor.

Applications

1. Transistors operating in a switching mode (class-C) appear to be reverse biased, or at least not fully forward biased. Class-B operation may also produce this effect, depending on the amplitude of the driving signal and the relative impedances of the signal source and input circuit.

2. The horizontal-sweep output and driver circuits used in transistor tv receivers appear to be reverse biased because they are operated as switches (alternately driven into saturation and cutoff). The output stage and most driver stages are, in fact, driven into zener conduction during the cutoff cycle. The emitter-base diode conducts backwards during cutoff. Driving a transistor in this manner sweeps charge carriers out of the base region, assuring a complete and faster cutoff. One of the causes of horizontal-output transistor failure is an improper drive signal. A driving waveform of improper shape or amplitude will often cause the output transistor to overheat.

3. Many oscillator circuits operate in a switching mode and appear to be reverse biased. Blocking oscillators operate class-C and there is a reverse-bias dc voltage on the base when the circuit is oscillating. This, in fact, is one means of determining if the circuit is oscillating by simple dc voltage measurements.

4. The bias for some circuits is derived from signal rectification. An example is the sync separator circuit used in most tv receivers. The transistor is reverse biased except during sync time. The reverse bias is developed from and is proportional to the amplitude of the signal applied to the base. Such circuits are usually operated with the emitter grounded, producing maximum negative base potential.

REVIEW

Questions

Examine the schematics in Fig. I-2. All transistors and diodes are silicon types. Assume an audio sine wave of sufficient amplitude to cause diode switching.

33

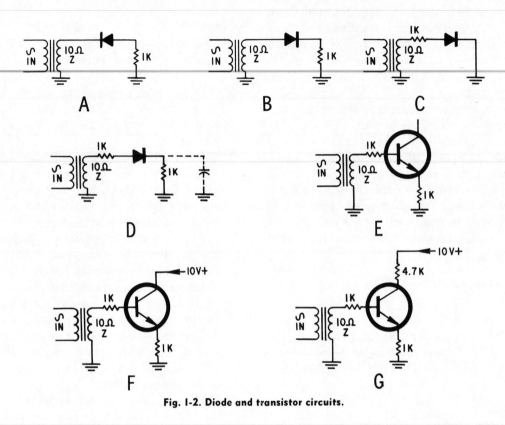

Fig. I-2. Diode and transistor circuits.

Q1. Describe the signals and the polarity of the dc voltages present at each end of the diode in Fig. I-2A. (Anode and cathode in relation to ground.)

Q2. Repeat this procedure for Fig. I-2B.

Q3. Repeat this procedure for Fig. I-2D (capacitor not connected).

Q4. What are the dc polarities in Fig. I-2C?

Q5. How will connecting a large-value capacitor as shown in Fig. I-2D affect the dc voltages present at the anode and cathode? Why?

Q6. Will the ac and dc conditions of Fig. I-2D and I-2E be identical? (Assume the capacitor in Fig. I-2D is not connected.)

Q7. Will the ac and dc conditions of the input circuit be identical in Figs. I-2E and I-2F?

Q8. The base voltage relative to ground in Fig. I-2G is −1 volt. Is this voltage a result of signal rectification by the emitter-base diode or saturation of the transistor?

Q9. Would connecting a large-value capacitor across the emitter resistors in Figs I-2F and I-2G affect the base voltages? How and why?

34

Answers

A1. Conduction occurs only during the negative excursion of the sine wave in Fig. I-2A. A negative dc voltage will appear at the anode as a result of the IR drop across the 1k resistor. The signal appears as the negative-going half of the ac sine wave. The dc voltage at the cathode will be essentially zero since the IR drop across the low resistance of the source winding is insignificant. Another way of saying the same thing is that the loading effect of the 1k resistor on the low resistance source winding cannot produce any appreciable dc voltage drop across the winding. The signal appearing at the cathode will be an ac sine wave with equal negative and positive excursions.

A2. Conduction occurs only during the positive excursion of the sine wave in Fig. I-2B. A positive dc voltage will appear at the cathode as a result of the IR drop across the 1k resistor. The signal is a positive-going half sine wave. All other dc and signal conditions will be as described in A1.

A3. Conduction occurs only during the positive excursion of the ac sine wave. A positive dc voltage will appear at the cathode as a result of the IR drop across the 1k resistor in the cathode circuit. The signal at this point will be a positive-going half wave. A negative dc voltage will appear at the anode as a result of the IR drop across the 1k resistor in the anode circuit. The two voltages will be approximately equal with opposite polarities. The signal present at the anode will be an ac sine wave with the negative excursion greater than the positive excursion. Another way of visualizing this is to consider that the anode circuit is loaded by the cathode circuit. The loading can only occur during conduction time, therefore, the positive half cycle is being loaded or reduced in amplitude while the negative half of the cycle is not affected at all. This may seem a very tedious means of expressing a rather simple concept. The intent of this approach is to develop thought patterns concerned with relative impedances developing various voltage (IR) drops, since this is really what transistor circuitry is all about.

A4. The anode is negative as a result of the voltage drop across the 1k resistor during the conduction cycle. The cathode voltage will, of course, be zero.

A5. The capacitor will cause both the anode and cathode voltages to increase. The cathode voltage becomes more positive because the capacitor acts as a battery or reservoir, tending to maintain the positive voltage between conduction cycles. The average measured dc voltage is, therefore, greater. The anode voltage becomes more negative because the capacitor acts as a low impedance, causing increased loading of the 1k resistor in the anode circuit. Therefore, more current flows during the conduction cycle causing an increased IR drop across the resistor.

A6. Yes. The electrical characteristics of silicon diodes are very similar when operating at the frequencies and power levels indicated. The only parameter which may cause different readings between the two circuits is the zener point. The zener voltage of most silicon emitter-base diodes is from 7 to 10 volts. In the circuit of Fig. 1-2E, this would limit the dc voltage developed at high signal levels.

A7. Definitely not. In the circuit of Fig. I-2F the collector is connected to a positive supply voltage. The device is operating as a transistor and the collector source is supplying the majority of the current flowing through the emitter resistor. The IR drop across this resistor now produces an impedance to the base circuit which is approximately equivalent to the current gain (beta) times the value of this resistor. If we assume a beta of 100, the base would be looking into an impedance of 100k ohms. The loading effect this would have on the 1k resistor in the base circuit would be negligible, and the dc voltage produced at the base by signal rectification would be about zero.

A8. Saturation. As explained in A7, the base voltage produced by signal rectification would be almost zero because of the relatively high impedance of the input circuit compared to the signal source. When saturation occurs, the collector-base diode becomes forward biased and shunts out the high-impedance emitter-base circuit. The collector-base diode now loads the 1k resistor in the base circuit during conduction cycles producing the negative 1 volt. Another way of saying the same thing is that signal rectification now takes place in the collector-base diode.

A9. Yes. The base voltage will become more negative in each case. Some negative voltage will be developed in the circuit in Fig. I-2F due to the lower ac impedance of the input circuit. Several actions occur in the circuit of Fig. I-2G. The input impedance is lowered as in circuit I-2F, producing some signal rectification by the emitter-base diode. Saturation is also occurring at a much lower signal level since the gain of the circuit is greatly increased by addition of the capacitor. (Any signal appearing across the emitter resistor is degenerative because of its polarity. A signal across the emitter resistor has the same effect as a reverse dc voltage across the emitter resistor which raises the input impedance of the circuit.)

J. Transistor Biasing Determines the Class of Operation and Stabilizes the Operating Point of the Transistor

Transistors, in common with most active electronic devices, can be forward biased, reverse biased, or zero biased. The polarity and level of bias determines the class of operation: class-A, class-B, or class-C.

When biased class-A, the degree of bias, in conjunction with the collector load resistor and the beta of the transistor, determines the operating point (the so called "Q" point) of the transistor. Simply stated, the "Q" point can be considered as the ratio of supply voltage to the voltage across the load resistor (or across the transistor). Beta

is a highly variable parameter even among transistors of the identical type number. A further complication is that beta changes with temperature, generally increasing as temperature increases. Since the operating point is a function of beta, a means of stabilizing this point must be achieved in a practical amplifier circuit, particularly if it is operating near its maximum power-dissipation capability.

Increasing temperature can also create another problem as a result of collector-to-base leakage. As the temperature rises, this leakage becomes greater, acting as an increase in forward bias. The collector-emitter current therefore rises, causing yet higher temperature. Such an unstable condition not only changes the operating point, but can also damage the device if it is operating near its maximum power-dissipation capability (*thermal runaway*). Practical biasing circuits must incorporate feedback that automatically compensates for the inherent instability of transistors due to the effects of beta and temperature. Let us first consider the various classes of operation.

Forward Bias

A transistor is forward biased when a potential is applied to the base in excess of the internal voltage drop across the base-emitter diode. The diode drop is about 0.5 to 0.6 volt for silicon transistors. As the forward-bias voltage applied to the base is increased, the resistance between emitter and collector decreases. Forward bias is the most common mode of operation since both positive and negative excursions of the signal are amplified, providing the transistor is not biased into saturation and the signal does not overcome the dc bias conditions. This is known as class-A operation.

Zero Bias

If no bias potential is applied to the base or the base is at the emitter potential, the transistor will not conduct. However, an ac signal of sufficient amplitude applied to the base of the transistor will cause conduction on each half cycle. A npn transistor will conduct on the positive half cycle and a pnp transistor will conduct on the negative half cycle. Conduction does not take place until the signal exceeds the internal voltage drop of the base-emitter diode. Because of this, the output signal would not be a faithful reproduction of the input signal. This type of distortion is referred to as *crossover distortion* and occurs in class-B amplifier stages that are insufficiently forward biased. True class-B operation would require a bias exactly equal to the internal voltage drop of the base-emitter diode. Normally, this is not practical since this voltage varies somewhat with temperature and from one transistor to another. Rather, a forward bias is applied to eliminate the crossover distortion and most so-called class-B transistor amplifiers are actually operated class-AB.

Reverse Bias

A potential can be applied to the base which reverse biases the emitter-base diode. The transistor will not conduct until a signal of sufficient amplitude is applied to overcome the reverse bias and forward bias the transistor. This is class-C operation. Depending on the level of reverse bias, any portion of a given input signal can be selected for amplification. As stated previously, reverse bias can also be developed by signal rectification. Sync-separator circuits used in most tv receivers develop reverse bias in this manner. Reverse-biased circuits are also useful as level-detector circuits where a specific change in a dc voltage must be sensed. An example of such a circuit is the color-killer circuit used in many color tv receivers.

Bias and Stability

The simplest means of forward biasing a transistor is to connect a suitable resistance from one side of the power supply to the base, observing correct polarity to cause conduction. (See Fig. J-1.) Biasing in this manner does not provide any stability since the bias value is fixed by the resistance. This type of biasing is suitable only if the transistor is operated in and out of saturation, such as a switch.

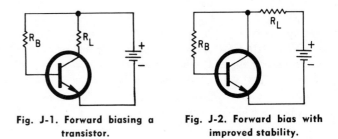

Fig. J-1. Forward biasing a transistor.   Fig. J-2. Forward bias with improved stability.

Improved stability is achieved by connecting the bias resistor to the opposite side of the collector load resistor as shown in Fig. J-2. Increased transistor conduction results in a greater voltage drop across the collector load resistor, reducing the bias applied to the base. This form of biasing is sometimes used but has the disadvantage of also feeding the signal across the load resistor back to the base. This signal is of a polarity which causes degeneration or loss of gain.

The most common means of achieving stability is to introduce negative or inverse feedback by placing a resistor in the emitter circuit as shown in Fig. J-3. Any increase in output current due to a temperature or bias change results in an increased voltage drop across the emitter resistor. The polarity at the emitter side of the resistor (Fig. J-3) becomes more positive (npn transistor), reducing the forward bias and automatically restricting the output current. The signal appearing across this resistor also causes degeneration or loss of gain, as did the signal across the collector load resistor. However, in the common-emitter circuit configuration, the output signal is taken off at the collector. Therefore, the emitter resistor can be bypassed with a capacitor, effectively shorting the ac signal to ground and preventing degeneration.

The emitter resistor, in conjunction with beta, also determines the dc resistance and ac impedance (when not bypassed) of the input circuit. This is a direct result of the inverse-feedback voltage developed across the emitter resistor and approximates beta times the value of the emitter resistor.

Feedback produced by emitter bias is seldom used as shown, since the effects of collector-to-base leakage will still result in a circuit of poor stability. The collector-to-base leakage, or reverse current, has only one possible path as shown in Fig. J-3, and that is through the emitter-base (control) circuit. The effect is an increase in bias current and a shift in the operating point. The effects of leakage current can be reduced by supplying the base bias from a divider network as shown in Fig. J-4.

The collector-to-base leakage current now has an alternate or parallel path through RB_2. It can be readily seen that the lower RB_2 becomes in value, the more stable the circuit becomes since less leakage current flows through the emitter-base circuit. The stability factor or *S factor* is, in fact, approximately equal to the value of the RB_2 resistance divided by the emitter resistance.

$$S = \frac{RB_2}{R_E}$$

If we assign arbitrary values to these components, it becomes apparent why this is so. Let us assume that R_E equals 1k and RB_2 is equal to 20k. We will also assume the transistor beta to be 100. The stability factor in this case would be about 20.

$$S = \frac{RB_2}{R_E}$$
$$= \frac{20k}{1k}$$
$$= 20.$$

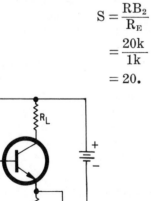

Fig. J-3. Using inverse feedback to achieve stability.

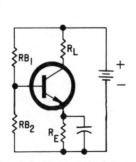

Fig. J-4. Bias supplied from divider network.

The lower the S factor is the greater the stability. Dividing the beta by the S factor, we can determine that the effect of leakage current has been reduced by a factor of five. This makes sense, since RB_2 is really 5 times lower in value than the impedance of the transistor input circuit, which is beta times R_E.

$$100 \times 1000 = 100,000 \text{ ohms}$$

As a result, about 5 parts of the total leakage current is flowing through RB_2 and only 1 part through the emitter-base circuit.

The stability factor required depends a great deal on the particular circuit, the transistor beta, and transistor construction. Silicon transistors, having lower leakage, do not require as low an S factor as germanium transistors. Low-power circuits do not require the stability of high-power circuits. High stability is usually obtained at the sacrifice of gain since RB_2 loads the base circuit. For this reason, circuits are designed to have high stability only when it is actually required. A stability factor of 20 is very good. If lower S factors are required, combination techniques are sometimes used, such as employing both collector feedback and emitter feedback.

High-power circuits, such as audio-output stages, often require S factors below 5 and usually employ other devices in addition to resistive stabilization methods. Diodes are frequently used to develop a bias voltage which changes with temperature. Many other temperature-sensitive devices are also employed by various manufacturers. Heat is often minimized by mounting the output transistors on metal heat sinks.

The methods just described all involve single stages. Practical amplifiers are usually multistaged and, consequently, overall feedback techniques can be used to stabilize the operating point. A common practice, particularly in direct-coupled audio amplifiers, is to feed a portion of the output signal back several stages to the input. This negative feedback can be either a dc voltage change, an ac signal, or both.

Applications

1. Transistor conduction depends on the base voltage and the value of the emitter resistor. In many circuits, the voltage appearing across the emitter resistor can be a useful indication of transistor operation.
2. In many circuits, the emitter resistor is the most important and critical component since it virtually controls the degree of conduction. Many high-power circuits have very low value

emitter resistors. Typical values are 0.39 ohm, 0.47 ohm, 1 ohm, etc. Care should be taken to replace these resistors with components of identical value.

3. The emitter bypass capacitor can greatly affect the stage gain. Open or low-value capacitors should always be suspected when low gain is encountered. An amplifier can also be made somewhat frequency-selective by choosing appropriate values of emitter bypass capacitance.

4. When replacing transistors or temperature-sensing devices which are mounted on heat sinks, be careful to fasten securely and use heat-conducting greases where required. Some transistors have heat sinks which merely are slipped over the transistor body. Do not forget to reinstall these heat sinks after replacing the transistor.

5. A good understanding of which components control stability can be quite useful when troubleshooting high-power audio stages. Distortion, in many cases, can be due to poor stability resulting from a change in component value or a component failure.

REVIEW

Questions

Examine the schematics in Fig. J-5.

Q1. Which circuit in Figs. J-5A through D has the highest output current?

Q2. All the voltages in Fig. J-5D measure correctly, yet the gain is low. What is the most likely problem?

Q3. A problem exists in one of the circuits. Which circuit is it and what is the problem?

Q4. What can occur if C1 in Fig. J-5D shorts?

Q5. Does C1 affect the frequency response of the amplifier in Fig. J-5D? In what manner?

Q6. How much output current is flowing from emitter to collector in Fig. J-5D?

Q7. Is the transistor in Fig. J-5E a practical circuit from a stability standpoint?

Q8. What is the approximate stability factor of the circuit in Fig. J-5F? Is such a stability factor required?

Q9. What is the approximate stability factor of the circuit in Fig. J-5G? Is such a stability factor required?

Q10. If the supply voltage in Fig. J-5G were reduced to 10 volts, would the same stability be required?

Q11. Assume all resistors are increased in value 10× in Fig. J-5G. If the same transistor is used, is the same stability factor required?

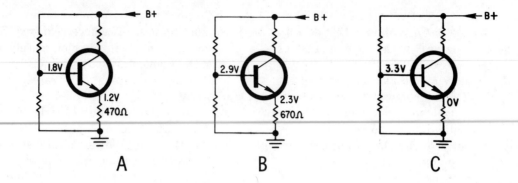

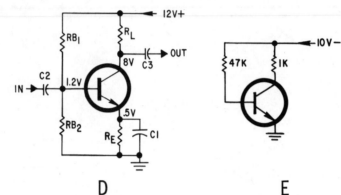

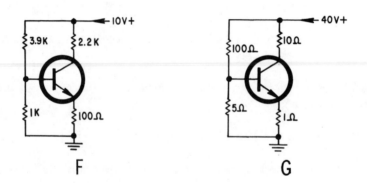

Fig. J-5. Typical transistor circuits.

Answers

A1. The transistor in Fig. J-5B is conducting the greatest. This can be simply calculated by dividing the voltage across the emitter resistor by the value of the emitter resistor:

$$I = \frac{E}{R}$$

In practical circuits, the collector current can be assumed to approximate the emitter current.

A2. Capacitor C1 is probably open or low in value. The signal appearing across the emitter resistor is degenerative and, unless bypassed to ground, reduces the gain of this stage.

A3. The problem exists in the circuit of Fig. J-5C. The transistor is not conducting as evidenced by the lack of voltage across the emitter resistor. The transistor is obviously forward biased due to the 3.3 volts on the base. The problem must be an open emitter-base diode.

A4. In most practical circuits, the transistor saturates if the emitter resistor is shorted. The output current will then depend on the value of the collector load resister R_L. In high-power circuits, R_L may be quite low, causing the maximum collector current to be exceeded and damaging the device. Even if saturation does not occur, stability will be lost and distortion will occur due to a change in the operating point.

A5. C1 will affect the frequency response. As C1 is made larger, the more-low frequencies will be bypassed to ground and the low-frequency response will be extended in range. If C1 is relatively low in capacity, only the higher frequencies are bypassed, resulting in poor low-frequency response.

A6. The output current cannot be determined. Transistor conduction depends on the base voltage and emitter resistor value. The emitter resistance is not given. The voltages given here only indicate that the transistor is forward biased and operating class-A.

A7. Certainly. The transistor is saturated and does not require stabilization. The resistor values shown would allow about a 50-to-1 ratio of output to input current. We assumed a beta of 100, or a 100-to-1 ratio. The transistor, therefore, must be saturated and the output current is limited by the 1k resistor.

A8. The stability factor is about 10.

$$S = \frac{1000}{100}$$
$$= 10$$

No, this stability factor is not required. Calculation will again show that the transistor is being operated as a switch (saturated). Stability, in this case, is not a factor since the current is determined primarily by the 2.2k collector load resistor. The biasing circuit could be redesigned, increasing the switching sensitivity. If the 1k resistor were, in fact, to be removed and the 3.9k resistor increased to about 100k, the transistor would still achieve saturation.

A9. The stability factor is about 5.

$$S = \frac{5}{1}$$
$$= 5$$

This degree of stability is no doubt required for this circuit, since simple calculation will show that the transistor is operating class-A and dissipating about 35 watts continuously. Most amplifiers incorporating high-power class-A output stages, such as car radios, utilize overall (multistage) feedback to stabilize the operating point.

A10. No, the same stability would not be required. In fact, with the values shown and no signal input, the transistor is cut off. If a signal were applied, approximately class-B operation would occur and a high degree of stability would again be desirable. However, the required stability would not be as high as required for class-A operation since less overall power is dissipated.

A11. No, the same stability factor would not be required. The particular transistor used in Fig. J-5G must be capable of dissipating at least 35 watts as indicated in A9. Increasing all resistors by a factor of 10 would require that the transistor dissipate only 3.5 watts. Operating the transistor in this manner, far below its maximum power handling capabilities, would allow a higher stability-factor number.

K. A Transistor Amplifies Because It Couples a Low-Impedance Circuit to a High-Impedance Circuit at About the Same Current Level

An amplifier, as the term is generally used in electronics, denotes a device which is capable of power amplification. A transformer can amplify voltage or current at the expense of one or the other but is not an amplifier. A transistor produces the useful power gain as voltage gain, current gain, or a combination of both, at virtually no expense to either. This should not, of course, be thought of as something for nothing since the additional power is being supplied by the power source.

All forms of gain, be they voltage or current, in transistor circuits must be considered in terms of impedance (dc and ac). The original inventors of the device called it a *transistor* for good reason. The term is a combination of the phrase *transfer-resistor*. This is precisely what a transistor does. It transfers the effects of a change occurring in a low-impedance circuit to a relatively high-impedance circuit at the same current. Considerable power gain can thus be realized; power being I²R. Transistors can be connected in three basic circuit configurations depending on which gain/impedance characteristics are to be utilized. These various configurations affect the signal processing but do not change the basic operation of the transistor. The characteristics of the three circuit configurations are covered fully in the next statement.

Determining the exact power gain of a single-stage transistor amplifier can be a very complex problem. Multiple stages compound the complexity. Fortunately, the voltage, current, and power gains can be quickly approximated in most circuits simply by inspection of resistance values and using some basic arithmetic. Although the ability to accomplish this is of no great importance to the average technician, it provides a useful understanding of how the ability of a transistor to amplify is related to other circuit components.

Gain is probably more easily understood if thought of in dc terms. Since ac at any given instant is actually a change in dc level, the mechanisms producing gain are essentially the same for both. Let us first consider voltage gain.

Voltage Gain

Voltage gain in any transistor circuit configuration is determined simply by the ratio of input impedance to output impedance (ac and dc). This can readily be illustrated by an example. Let us consider the emitter-collector circuit of any transistor, assign arbitrary voltage and resistance values, and observe exactly how voltage gain occurs.

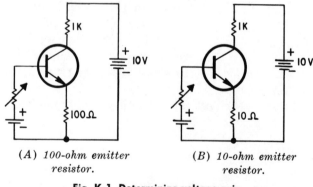

(A) 100-ohm emitter resistor.　　(B) 10-ohm emitter resistor.

Fig. K-1. Determining voltage gain.

Bear in mind that the emitter in all circuit configurations is part of the input circuit. When the transistor is forward biased, emitter voltage changes reflect a change in base voltage.

If we adjust the bias of the transistor in Fig. K-1A so that the emitter voltage is 0.5 volt, then by Ohm's law:

$$I = \frac{E}{R}$$
$$= \frac{0.5}{100}$$
$$= .005 \text{ ampere.}$$

The collector current in most practical circuits will approximate the emitter current, so the voltage across the load resistor becomes:

$$E = IR$$
$$= .005 \times 1000$$
$$= 5 \text{ volts.}$$

Now, let us increase the bias so that 0.6 volt appears across the emitter resistor. By Ohm's law the emitter current is:

$$I = \frac{E}{R}$$
$$= \frac{0.6}{100}$$
$$= .006 \text{ ampere.}$$

The voltage across the load resistor is now:

$$E = IR$$
$$= .006 \times 1000$$
$$= 6 \text{ volts.}$$

The bias increase has produced an emitter voltage change of 0.1 volt. This has produced a collector voltage change of 1 volt. Obviously, the voltage gain is 10. The collector load resistor (1k) divided by the emitter resistor (100) is also 10. Apparently the voltage gain in our example can be determined simply by dividing the collector resistor by the emitter resistor.

Let us see if this holds true when we change the emitter resistor value. In Fig. K-1B, the emitter resistor is 10 ohms. To produce a collector current of .005 ampere the bias will have to be set so that 0.05 volt appears across the emitter resistor. The voltage across the load resistor will again be 5 volts.

$$.005 \times 1000 = 5 \text{ volts}$$

An increase in bias producing an emitter voltage of 0.06 volt will, by Ohm's law, result in 6 volts across the collector load resistor. A 0.01-volt change of emitter voltage has resulted in a 1-volt change of collector voltage, or a voltage gain of 100. Again, the collector resistor divided by the emitter resistor also equals the voltage gain.

We will pursue this concept of voltage gain further with additional examples, showing the effect of the collector load resistor. Before going on, however, we should introduce a factor we have been overlooking for simplicity of illustration. The true emitter resistor value is the emitter resistor plus the intrinsic resistance of the emitter-base diode, which depends on the emitter current. This resistance, for both germanium and silicon transistors, is about 30 ohms divided by the emitter current in milliamperes (Shockley's Relation). Normally, this is represented as the ac impedance of the transistor in most publications. While true, this also represents the dc resistance for small changes in dc bias. The true emitter resistance in the two examples given would, therefore, be 106 ohms and 16 ohms respectively.

The intrinsic resistance of the base-emitter diode is:

$$\frac{30}{5 \text{ mA}} = 6 \text{ ohms.}$$

It should be apparent that this resistance becomes an important factor only if the emitter resistor is relatively low in value, nonexistent, bypassed, or the bias is such that the emitter current is very low in value. Consider the schematic in Fig. K-2.

Operation in the center of the class-A amplifier mode dictates a bias setting which will produce about .005 ampere of emitter-collector current. This produces a collector voltage of 5 volts or half of the supply voltage. The voltage gain will now be approximately equal to the collector load resistor divided by the intrinsic emitter resistance. Again, the intrinsic emitter resistance is about 30 ohms divided by the emitter current expressed in milliamperes.

$$\frac{30}{5} = 6 \text{ ohms.}$$

The voltage gain (G_v) will be:

$$\frac{1000}{6} = 166.$$

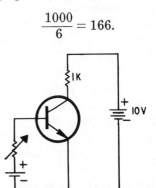

Fig. K-2. Determining voltage gain for a circuit without an emitter resistor.

When no emitter resistor is present, or if completely bypassed, a more accurate approximation of voltage gain is:

$$G_v = \frac{.9 \times R_L}{R_{INT}}.$$

Thus, in this case:

$$G_v = \frac{.9 \times 1000}{6}$$
$$= 150.$$

This is particularly true at higher values of emitter current.

Apparently, voltage gain is a function of the emitter resistance (internal and external) and the collector load resistor. Common sense would then lead one to assume that if the load resistor were 500 ohms, the voltage gain would be cut in half. This is indeed the case. However, the operating point will also change. Under these conditions, the voltage across the load resistor will be 2.5 volts.

$$E = IR$$
$$= .005 \times 500$$
$$= 2.5 \text{ volts.}$$

The gain would be:

$$G_v = \frac{R_L}{R_E}$$
$$= \frac{500}{6}$$
$$= 83.$$

If a change in operating point is not desirable, then the bias must be adjusted to maintain the same voltage gain. Let us consider how this occurs. If the collector load resistor in Fig. K-2 is 500 ohms, and operation in the middle of the class-A range is desired, a bias must be set up to produce an emitter-collector current of 0.01 ampere. The voltage across the load resistor will again be 5 volts $(0.01 \times 500 = 5)$. The intrinsic emitter resistance will be:

$$\frac{30}{10} = 3 \text{ ohms.}$$

The voltage gain will again be:

$$\frac{500}{3} = 166.$$

Increasing the value of the collector load resistor produces similar results in the opposite direction. If the resistance of the collector load resistor is doubled, one would expect the voltage gain to double, which it does up to the point of saturation. If the collector load resistor in Fig. K-2 were increased to 2000 ohms and the 0.005-ampere emitter-collector current is maintained, the collector voltage would be zero volts. All of the supply voltage is across the load and there is no voltage drop across the transistor, indicating saturation. The maximum theoretical voltage gain would occur just before saturation is reached. For the sake of illustration, we can say that maximum voltage gain is just at saturation as in our example. The voltage gain would be:

$$G_v = \frac{R_L}{R_E}$$
$$= \frac{2000}{6}$$
$$= 332.$$

An amplifier operating just below saturation would represent the greatest possible voltage gain, because the collector load resistance would be the largest resistance that can be used for that particular bias point. Of course, the limit of class-A signal swing under these conditions is at minimum.

Suppose we wished to operate the amplifier in Fig. K-2 at the class-A center point with a 2000-ohm load resistor. The bias must now be adjusted to cause .0025 amperes of emitter-collector current. The intrinsic emitter resistance will be:

$$\frac{30}{2.5} = 12 \text{ ohms.}$$

The voltage gain will again be:

$$G_v = \frac{R_L}{R_E}$$
$$= \frac{2000}{12}$$
$$= 166.$$

More accurately:

$$G_v = \frac{.9 \times R_L}{R_E}$$
$$= 150.$$

The preceding examples illustrate that maximum voltage gain is obtained when the external emitter resistor is zero and is independent of the collector load resistor when mid Class-A operation is desired. The preceding examples indicate that the greatest theoretical voltage gain, when operating in the middle of class-A mode with no emitter resistor, is 166. However, this is not the highest voltage gain possible. If the supply voltage were doubled to 20 volts, mid class-A operation could be achieved with an emitter-collector current of .005 ampere and a load resistor of 2000 ohms. The voltage gain would now be 2 times 166 or 332. An increase in supply voltage produces greater voltage gain because it allows the use of a larger load resistor. However, the load resistance divided by emitter resistance will still equal the voltage gain, regardless of supply voltage.

Does this voltage gain relationship apply to all three basic circuit configurations? Yes, but because of circuit arrangement, the common-base voltage gain equals the voltage gain of the common-emitter configuration, providing it is operating without feedback (no emitter resistor).

Notice the manner in which the signal is applied in the common-base configuration in Fig. K-3A. The signal input path is directly through the emitter-base junction and does not include the emitter resistor. The voltage gain will, therefore, be the load resistor divided by the intrinsic emitter resistance. The same conditions apply to the common-emitter circuit without an emitter resistor shown in Fig. K-3B and as illustrated in our previous examples.

Applying this concept of voltage gain to the common-collector circuit in Fig. K-3C, we discover that the voltage gain is always less than one. The

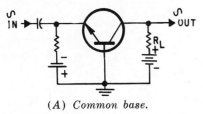

(A) Common base.

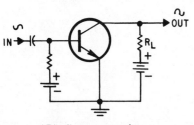

(B) Common emitter.

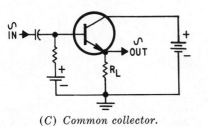

(C) Common collector.

Fig. K-3. Three basic transistor circuit configurations.

emitter resistor in this case is also the load resistance. The true emitter resistance is, again, the intrinsic emitter resistance plus the external emitter resistor. The total emitter resistance, as a result, is always larger than the load resistor by a quantity equal to the intrinsic emitter resistance. Therefore, at high emitter currents, the voltage gain of the common-collector circuit approaches one. The lower the emitter current, the greater the intrinsic emitter resistance becomes, resulting in less voltage gain. However, for all practical purposes, the voltage gain of the emitter-follower amplifier can be assumed as unity (one).

We have demonstrated that the voltage gains of the common-emitter and common-collector circuits are approximately determined by dividing the collector load resistor by the total emitter resistance (intrinsic and external). The voltage gain of the common-base circuit is approximately the collector load resistor divided by the intrinsic emitter resistance.

Why is the voltage gain approximate? One assumption made at the start of this chapter was that the collector and emitter currents were equal. All of the voltage-gain calculations were made assuming this equality. The collector current, in reality, is the emitter current less the base current. The base current, in turn, is dependent on beta. The higher the beta, the closer will be the voltage-gain approximation. This is because a higher beta will require less base current for a given collector current.

The theoretical voltage gain of a single stage is a simple matter of observing resistor values and applying basic arithmetic. This, unfortunately, does not tell the entire story regarding gain. Up to this point, we have described only voltage gain. In the final analysis, all useful gain must be in terms of power gain. Power gain can be described as the voltage gain times the current gain. Another consideration is that amplifiers are generally multistaged devices. Remember that in RC-coupled amplifiers, the total input impedance of one stage is in parallel with the output resistor of the preceding stage. The collector load resistor is effectively reduced in value and, in accordance with our voltage-gain formula, the voltage gain will be reduced. A further consideration is that most of the examples of voltage gain discussed thus far describe the maximum gain, or gain without feedback stabilization. Practical amplifiers cannot be designed in this manner because the operating point will change. The relation of feedback stabilization to voltage and current gain is explained subsequently under current gain.

Current Gain

The current gain of the common-emitter and common-collector circuits is simply beta. The current gain of the common-base circuit is alpha. The input of the common-base circuit is the emitter and the output is the collector. Since the collector current is always less than the emitter current, alpha is always less than one. If either beta or alpha is known, the other can be determined as follows:

$$\text{beta} = \frac{\text{alpha}}{1 - \text{alpha}}$$

$$\text{alpha} = \frac{\text{beta}}{1 + \text{beta}}$$

The power-gain capabilities of the common-base and common-collector amplifiers are less than the common-emitter configuration. The common-emitter circuit embodies both current and voltage gain. As has been demonstrated, the current gain of the common-base circuit is less than one and the voltage gain of the common-collector circuit is less than one.

Beta (and consequently alpha) is, as previously stated, a highly variable function. Practical amplifiers must be designed to function at least partially independent of beta. Practical amplifiers are seldom operated at maximum gain. Feedback is incorporated and gain is traded for stability. The current gain of an adequately stabilized common-emitter or common-collector amplifier is usually determined by the circuit resistors (Fig. K-4). The current gain G_I is about equal to the emitter-base resistor divided by the emitter resistor.

$$G_I = \frac{RB_I}{R_E}$$

Does this formula look familiar? It should. It is the same as the stability formula described in the preceding statement. The stability factor and the current gain are practically synonymous. Increasing R_E or decreasing RB_I creates greater stability at the expense of voltage gain and current gain respectively.

We usually cannot operate the common-base amplifier near its highest voltage-gain capability.

The input impedance is so very low that shunting losses and coupling problems occur. The emitter-base diode characteristics are also nonlinear due to zero feedback. To correct these problems, a low-value resistor is usually connected between the signal source and the emitter. This resistor must now be added to the intrinsic emitter resistance, raising the impedance and reducing the voltage gain.

Gain, up to this point, has been described in terms of small dc changes for the sake of clarity. We have observed the mechanisms involved in producing gain and concluded that the dc voltage and current gains must be partially sacrificed to prevent changes in the operating point. In some cases, however, we need not reduce the ac gain as drastically.

The operating parameters of the circuit in Fig. K-5 can be approximated by simple inspection of resistor values. Voltage gain is about 12 (2500 divided by 200). Current gain and stability factor is about 11 (2200 divided by 200). The intrinsic emitter resistance is about 7 ohms (30 divided by 4 mA). Consider now the same circuit with the emitter resistor bypassed (Fig. K-6). The emitter is considered fully bypassed if the capacitive reactance of the capacitor is lower than the intrinsic emitter resistance. In this case, the capacitive reactance of 1000 μF at 100 hertz is less than 2 ohms.

$$X_C = \frac{1}{2\pi fC}$$
$$= \frac{1}{6.28 \times 10^2 \times 10^{-3}}$$
$$= 1.6 \text{ ohms.}$$

The voltage gain is now a function of the intrinsic emitter resistance and is greater than 300.

$$G_v = \frac{.9R_L}{R_{INT}}$$
$$= \frac{.9 \times 2500}{7}$$
$$= 321.$$

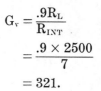

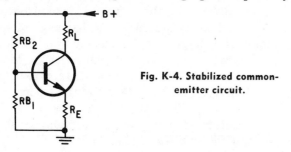

Fig. K-4. Stabilized common-emitter circuit.

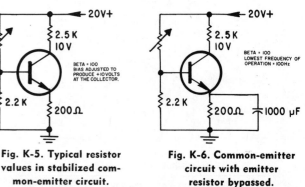

Fig. K-5. Typical resistor values in stabilized common-emitter circuit.

Fig. K-6. Common-emitter circuit with emitter resistor bypassed.

Circuits are seldom fully bypassed for several reasons. Remember that the output impedance of the previous stage is shunted by the input impedance of the next stage. A very low input impedance limits the gain of the previous stage. Complete bypassing also necessitates prohibitively large bypass and coupling capacitors. Complete bypassing also causes distortion because of the variable nature of the intrinsic emitter resistance. Since the emitter current changes with signal, the intrinsic emitter resistance also changes and shows up as distortion. A common practice is to connect a low-value, unbypassed resistor (usually less than 100 ohms) in series with the emitter resistor as shown in Fig. K-7.

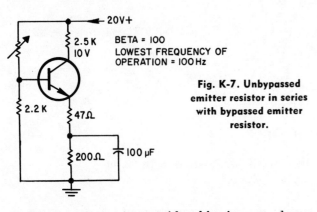

Fig. K-7. Unbypassed emitter resistor in series with bypassed emitter resistor.

This amplifier is considerably improved over that shown in Fig. K-6. The input impedance has been raised from about 700 to about 7000 ohms. (The total input impedance being 47 ohms + 7-ohms intrinsic resistance + 16-ohms capacitive reactance; all times beta.) The bypass capacitor has been reduced by a factor of 10. Adequate bypass-

ing occurs so long as the capacitive reactance at the desired frequency is considerably less in value than the unbypassed resistor. The ac-voltage gain is about 36.

$$\frac{2500}{47 + 7 + 16} = 35.7.$$

The ac current gain is about 31.

$$\frac{2200}{70} = 31.4.$$

Applications

1. A means of quickly estimating voltage gain is quite useful when troubleshooting circuits suspected of being low in output.
2. The emitter bypass capacitor greatly affects stage gain. Open or low-value capacitors should always be suspected when low gain is encountered.
3. Thinking of gain in terms of relative impedances is very useful when signal tracing with a scope or ac voltmeter. The signal amplitudes appearing at the various transistor elements and in various parts of the circuit can be anticipated. Signal levels which deviate widely from expected levels can be pinpointed as problem areas. A simple example would be an open emitter-bypass capacitor. Normally, little or no signal appears at the emitter resistor if it is bypassed. However, the presence of a considerable signal at this point would indicate a defective capacitor.
4. The ability to recognize and anticipate the gains of the three basic circuit configurations is useful when troubleshooting.

REVIEW

Questions

Examine the schematics in Fig. K-8.
Consider operation of both amplifiers at 100 Hz.
All transistors are silicon types with a beta of 100.

Q1. List the approximate voltages appearing at each of the transistor elements in the schematics shown in Fig. K-8.

Q2. Which amplifier has the greatest overall voltage gain, Fig. K-8A or Fig. K-8B?

Q3. What is the approximate overall voltage gain of each amplifier?

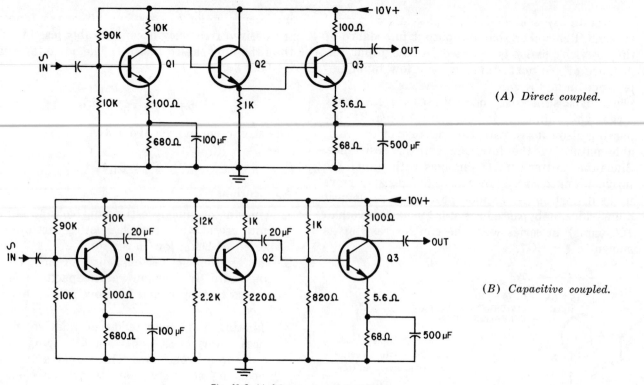

Fig. K-8. Multistage transistor circuits.

Q4. Are the coupling capacitors used in Fig. K-8B large enough in value for operation at 100 Hz?

Q5. Fig. K-9 is a partial schematic of an amplifier used for remote-control applications in color tv receivers. Assume operation at 35 kHz. All transistors are silicon. Beta = 100. Calculate the maximum and minimum overall voltage gain.

Answers

A1. *Fig. K-8A*

Q1	E	.4 Volt
	B	1.0 Volt
	C	5.1 Volts
Q2	E	4.5 Volts
	B	5.1 Volts
	C	10.0 Volts
Q3	E	3.9 Volts
	B	4.5 Volts
	C	5.3 Volts

Fig. K-8B

Q1	E	.4 Volt
	B	1.0 Volt
	C	5.1 Volts
Q2	E	0.9 Volt
	B	1.5 Volts
	C	4.1 Volts
Q3	E	3.9 Volts
	B	4.5 Volts
	C	5.3 Volts

A2. Fig. K-8A has the greatest voltage gain.

A3. The total voltage gain of the amplifier in Fig. K-8A is about 530. The amplifier in Fig. K-8B has a a voltage gain of about 90. The component values chosen for the amplifier in Fig. K-8B were deliberately selected to illustrate the effects of emitter feedback and impedance loading on amplifier gain. The step-by-step derivation of these gain figures is as follows:

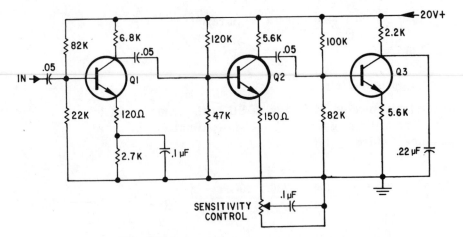

Fig. K-9. RC-coupled stages.

Fig. K-8A

The gain of Q3 approximates 100 (collector load) divided by 5.6 + 3 (the capacitive reactance of 500 µF at 100 Hz) +0.6 (the intrinsic emitter resistance since the emitter current by Ohm's law is about 50 mA). Therefore, the voltage gain of Q3 is approximately 11.

$$G_v = \frac{100}{5.6 + 0.6 + 3}$$
$$= 10.9.$$

The gain of Q2, an emitter follower, is 1.

The gain of Q1 is 8300 (total collector load) divided by 100 + 15 (the capacitive reactance of 100 µF at 100 Hz) + 60 (the intrinsic emitter resistance since the emitter current by Ohm's law is about 0.5 mA).

The total collector load impedance of Q1 approximates the input impedance of Q2 paralled with the 10k collector load resistor. The impedance of Q3 is 1k (beta × 10). This is in parallel with the 1k emitter resistor of Q2 which gives a total emitter resistance of 500 ohms for Q2. Therefore, the input impedance of Q2 is 50k (500 × beta). The 10k collector load resistor in parallel with the 50k input impedance of Q2 gives a total collector load impedance of 8300 ohms for Q1. Therefore, the voltage gain Q1 is about 48.

$$G_v = \frac{8300}{100 + 60 + 15}$$
$$= 47.5.$$

The total voltage gain of the amplifier in Fig. K-8A is approximately 530.

$$G_v = 48 \times 1 \times 11$$
$$= 528.$$

Fig. K-8B

The gain of Q3 in Fig. K-8B is the same as Q3 in Fig. K-8A (11) and is derived in the same manner.

The voltage gain of Q2 is the total collector impedance divided by 227 (the emitter resistor plus the intrinsic emitter resistance of 7 ohms). The total collector load impedance approximates the parallel impedance of Q3, the two base resistors, and the collector resistor of Q2.

$$G_v = \frac{1000 \parallel 1000 \parallel 820 \parallel 1000}{227} \quad (\parallel = \text{parallel})$$

$$= \frac{235}{227}$$

$$= 1 \text{ approximately.}$$

The total emitter resistance of Q1 in Fig. K-8B is the same as Q1 in Fig. K-8A (175). The voltage gain of Q1 is approximately the total collector impedance divided by 175.

Therefore:

$$G_v = \frac{22000 \parallel 12000 \parallel 2200 \parallel 10000}{175}$$

$$= \frac{1460}{175}$$

$$= 8 \text{ approximately.}$$

The total voltage gain of the amplifier in Fig. K-8B is approximately 90.

$$G_v = 8 \times 1 \times 11$$
$$= 88.$$

A4. Yes, the impedance of 20 μF at 100 Hz is about 80 ohms. Low-loss coupling is obtained if the impedance of the capacitor at the desired operating frequency is considerably less than the input impedance of the circuit into which it couples. 20 μF is really much larger than required, since the coupling is into circuits of 1700 ohms and 310 ohms respectively.

The input impedance of Q2 is 22,000 ohms (beta $\times$ 220). The total input impedance of Q2 as seen by the coupling capacitor is 1700 ohms.

$$22000 \parallel 12000 \parallel 2200 = 1700.$$

The total impedance of Q3 is 1000 ohms (beta $\times$ 10). The total input impedance of Q3 as seen by the coupling capacitor is 310 ohms.

$$1000 \parallel 820 \parallel 1000 = 310.$$

A5. The maximum overall voltage gain is approximately 552. The minimum 48. This is derived as follows: The voltage gain of Q3, emitter follower, is about 1. The voltage gain of Q2 is approximately the total impedance of the collector load divided by the total emitter impedance.

Therefore, at maximum gain setting the approximate voltage gain is 23.

$$G_v = \frac{100,000 \parallel 560,000 \parallel 82,000 \parallel 5600}{150 \, (R_E) + 45 \, (X_C) + 20 \, (R_{INT})}$$

$$= \frac{5000}{215}$$

$$= 23.$$

At minimum gain setting the approximate voltage gain is 1.6.

$$G_v = \frac{100,000 \parallel 560,000 \parallel 82,000 \parallel 5600}{150 + 3000 + 20}$$

$$= \frac{5000}{3170}$$

$$= 1.6.$$

The voltage gain of Q1 will also be somewhat affected by the setting of the sensitivity control in the emitter circuit of Q2, since the input impedance of Q2 changes with the control setting. The input impedance is beta times 215 (21,500) at maximum gain setting, and beta times 3170 (317,000) at minimum gain setting.

The approximate voltage gain of Q1 at minimum setting is 30.

$$G_v = \frac{317,000 \parallel 120,000 \parallel 47,000 \parallel 6800}{120\ (R_E) + 45\ (X_C) + 25\ (R_{INT})}$$
$$= \frac{5600}{190}$$
$$= 30.$$

The approximate voltage gain of Q1 at maximum gain setting is 24.

$$G_v = \frac{21,500 \parallel 120,000 \parallel 47,000 \parallel 6800}{120 + 45 + 25}$$
$$= \frac{4500}{190}$$
$$= 24.$$

The overall voltage gain is therefore:

Maximum 23 × 24 = 552
Minimum 1.6 × 30 = 48.

L. A Transistor Can Be Connected in Three Basic Circuit Configurations Each Having Different Characteristics

The three circuit configurations illustrated in Fig. L-1 are the common base, the common emitter, and the common collector. The transistor element which is common to both the input circuit and the output circuit denotes the particular circuit configuration.

A great deal of information has been written describing the impedance and gain characteristics of the three basic amplifier configurations. Unfortunately, most of these descriptions are highly mathematical in nature and tend to hide some simple concepts from the average technician not versed in this language. The lack of a good working knowledge in this area is responsible for most of the negative feeling many technicians have regarding solid-state electronics. A good understanding of the relative impedances associated with transistor circuitry does much to dispel the alleged mystery surrounding solid-state circuits. This area is generally confusing and, consequently, understanding it is of greatest benefit to the technician well versed in vacuum-tube technology.

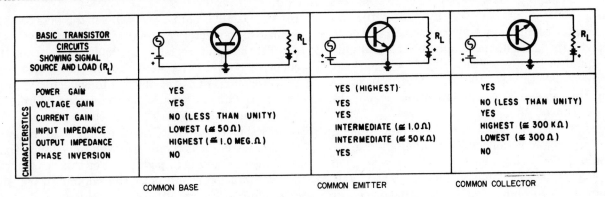

BASIC TRANSISTOR CIRCUITS SHOWING SIGNAL SOURCE AND LOAD (R$_L$)			
CHARACTERISTICS			
POWER GAIN	YES	YES (HIGHEST)	YES
VOLTAGE GAIN	YES	YES	NO (LESS THAN UNITY)
CURRENT GAIN	NO (LESS THAN UNITY)	YES	YES
INPUT IMPEDANCE	LOWEST (≅ 50 Ω)	INTERMEDIATE (≅ 1.0 Ω)	HIGHEST (≅ 300 KΩ)
OUTPUT IMPEDANCE	HIGHEST (≅ 1.0 MEG. Ω)	INTERMEDIATE (≅ 50 KΩ)	LOWEST (≅ 300 Ω)
PHASE INVERSION	NO	YES	NO
	COMMON BASE	COMMON EMITTER	COMMON COLLECTOR

Fig. L-1. Three basic transistor circuits.

Impedance, as the term is used here, includes both ac and dc resistance, providing dc resistance is considered in terms of a small change in dc level. When we consider the impedance characteristics of a transistor, we must visualize these factors as being load impedances or source impedances. The input terminals of a transistor are considered load impedances to an external source. The output terminals become a source impedance feeding an external load. See Fig. L-2.

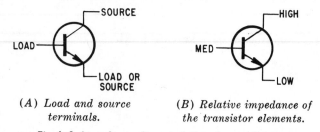

(A) *Load and source terminals.*   (B) *Relative impedance of the transistor elements.*

Fig. L-2. Impedance characteristics of a transistor:

There is nothing very mysterious about the relative impedances present at the three transistor elements. The load impedances are a result of the internal signal currents of the device. A signal taken off at the collector behaves as if generated by a high-impedance source due to the reverse-biased collector-base diode. The emitter behaves as a low-impedance source because of the current gain of the transistor.

Let us consider the loading or input impedance of a transistor as shown in Fig. L-3. At first glance, it may appear that the common-base and common-emitter configurations have identical input impedances. Such is not the case, since the common-emitter impedance is greater by a factor of beta. The reason for this is the action of the current from the collector supply on the internal emitter-base diode resistance, R_I.

The value of this internal resistance is approximately equal to 30 ohms divided by the emitter current in milliamperes. The signal current from the collector supply develops a voltage across this resistance, the polarity of which opposes the applied signal. Assuming an emitter current of 1 mA, the input impedance of the common-base configuration would be 30 ohms (Shockley's Relation). If the transistor beta were 100, the input impedance of the common-emitter configuration would be 100×30 or 3000 ohms.

Under these same conditions the input impedance of the common-collector amplifier in Fig. L-3C would be beta times $R_I + R_L$. If R_L were 1000 ohms, the input impedance would be $(30 + 1000) \times 100$ or 103,000 ohms. The input impedances of the common-emitter and common-collector ampli-

fiers are the same, providing the emitter resistor used equals the load resistor of the common-collector circuit.

The output impedance of both the common-base and common-emitter amplifiers is usually very high compared to the external circuit impedances. This can be readily demonstrated in terms of dc current flow. Assume we have a transistor operating from a 10-volt power supply. The collector load

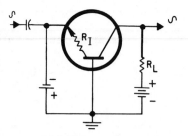

(A) *Common base.*

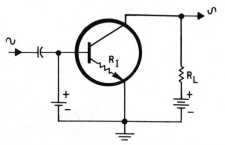

(B) *Common emitter.*

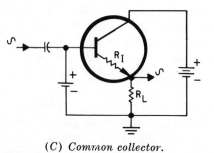

(C) *Common collector.*

Fig. L-3. Input impedances of three basic circuit configurations:

resistor is 1000 ohms. We will set the bias so that 1 mA of collector current flows. Now we will double the load resistor to 2000 ohms. Has the collector current changed? No, of course not. The collector current is insensitive to collector voltage changes. This is because the collector current is being generated by a high-impedance source, the reverse-biased diode. The external load resistor in most circuits is a much lower impedance than the collector source impedance. For this reason, the output impedance of both the common-base and common-emitter amplifiers is considered to be that of the load resistor.

The output impedance of the common-collector amplifier (also called an emitter follower) is an interesting study and may be rather surprising at first encounter. The output impedance of the amplifier shown in Fig. L-4 is about 70 ohms. Beta is 100. The load resistor of the common-collector amplifier is a factor in determining output impedance only because of its effect on emitter current. This, in turn, affects the intrinsic emitter-base resistance. The output impedance of the com-

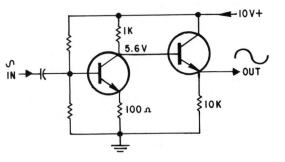

Fig. L-4. Common-collector circuit.

mon-collector amplifier is, in a manner of speaking, the reverse of its input impedance. We determined previously that the input impedance of the common-collector amplifier was beta times the load resistor plus the intrinsic emitter-base resistance. The output impedance is the intrinsic emitter-base resistance times beta plus the source impedance (1k), all divided by beta.

The emitter current is:

$$I = \frac{E}{R}$$
$$= \frac{5}{10,000}$$
$$= 0.5 \, mA.$$

The intrinsic emitter-base resistance is, therefore, 60 ohms. The output impedance then becomes:

$$Z_O = \frac{beta \times R_I + 1000}{beta}$$
$$= \frac{100 \times 60 + 1000}{100}$$
$$= \frac{6000 + 1000}{100}$$
$$= 70 \, ohms.$$

Applications

1. The common-emitter circuit is by far the most widely used circuit in home-entertainment electronics because it has both voltage and current gain.
2. The common-collector amplifier is the second most common and is used as a matching circuit. All of its gain appears as current gain.
3. The output and input signals of an emitter follower are virtually the same.
4. The ability to estimate various impedance levels is useful when troubleshooting since gain and amplitude levels in various portions of the circuit can be anticipated.

REVIEW

Questions

Examine the schematics in Fig. L-5.
All transistors are silicon.
Beta = 100

Q1. What is the approximate input impedance of the amplifier in Fig. L-5A?

Q2. What is the approximate input impedance of the amplifier in Fig. L-5B?

Q3. How much voltage must be applied to the base in Fig. L-5C to cause saturation?

Q4. The base of transistor Q2 in Fig. L-5D reads 1.6 volts with respect to ground. What is the problem?

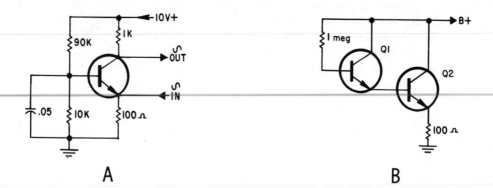

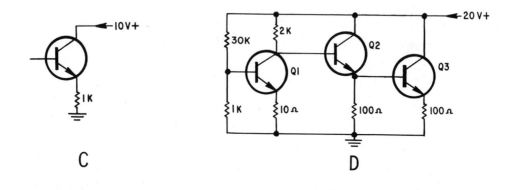

Fig. L-5. Typical transistor circuits.

Answers

A1. About 14 ohms. The intrinsic emitter-base impedance is about 7 ohms since by Ohm's law .004 ampere of emitter current is flowing. 30 ohms divided by 4 mA is then approximately 7 ohms. The capacitive reactance of .05 μF at 455 kHz is also about 7 ohms.

A2. About 500k ohms. The input impedance of Q2 is beta × 100 or 10,000. The input impedance of Q1 is beta × 10,000 or 1,000,000. 1 megohm and 1 megohm in parallel is 500k.

A3. The closest approximation of a saturated condition would be obtained by applying 10 volts to the base. We would now have a diode in series with the 1k resistor. Since an emitter follower does not have a collector load resistor, saturation does not occur at a lesser voltage.

A4. The collector of Q2 is open. Normally, the input impedance of Q2 is beta × 100 or 10,000 ohms. When Q2 opens we have essentially a 100-ohm resistor, a diode and a 2000-ohm resistor in series. The divider action of this combination plus the 0.6-volt diode drop produces 1.6 volts at the base. Let us consider what Q2 base voltage would be produced by other circuit malfunctions.
(a) If the emitter-base diode of Q2 were shorted, the input impedance would also be lost. However, we would not have the 0.6-volt drop and the base voltage would be 1 volt.

(b) If Q1 were saturated or shorted, the base voltage of Q2 would be about 0.1 volt due to the divider action of the 2000-ohm resistor and the 10-ohm resistor.

(c) What about Q3? If the collector were open or the emitter-base diode were shorted, the normal input impedance of 10,000 ohms would drop to about 100 ohms. Would this be low enough to load down the emitter and base of Q2? To answer this question we must know the output impedance of Q2. Q1 is biased to produce class-A operation. Ohm's law will show that the collector voltage will be about 12 volts. The emitter voltage of Q2 will, therefore, be 11.4 volts. The emitter current is then:

$$I = \frac{E}{R}$$
$$= \frac{11.4}{100}$$
$$= .1 \text{ ampere approximately.}$$

The intrinsic emitter base resistance is, therefore, 30 divided by 100 mA, or 0.3 ohm.

$$Z_o = \frac{.3 \times \text{beta} + \text{Rs}}{\text{beta}}$$
$$= \frac{30 + 2000}{100}$$
$$= 20 \text{ ohms approximately.}$$

The output impedance of Q2 is much lower than the 100-ohm emitter resistor of Q3. As a result, failure of Q3 cannot load Q2 sufficiently to drop the emitter and base voltage significantly.

PART II

CIRCUITS & PROCEDURES

CHAPTER 1

Test Equipment

Test equipment is specified for each circuit discussed. However, the equipment does not imply that these are the only instruments useful for troubleshooting the particular circuit. All or any of this equipment may not be required in every instance. Obviously, if an open connection can be located visually, an ohmmeter is not needed to track it down. In many cases, an ac meter can be substituted for an oscilloscope. The criteria in such cases must always be what is available and most convenient to use. Another important and often-overlooked factor is the ability and familiarity of the technician with the test equipment selected. Some very good technicians seldom use an oscilloscope. Experience enables them to accomplish as much with very simple equipment as others, less experienced, produce with more sophisticated instruments. Efficient technicians, particularly outside technicians, learn to bend the rules and apply shortcuts which are seldom recommended or condoned in service publications. The suggestion is not that indiscriminate "screwdriver-mechanic" techniques be applied to service, but rather that many simple jumpering and noise-injection practices are often the quickest means of isolating a problem.

The equipment specified here is included among the test instruments found in the average shop. In almost every case, a multimeter of some sort is indicated since this instrument is indispensable. An oscilloscope is generally specified since most shops have such an instrument and would probably benefit by using it more. The following paragraphs describe the equipment and service aids recommended and the uses of each.

JUMPER LEADS

Several jumper wires of various lengths fitted with miniature alligator clips at each end are a valuable service aid. A handy addition is a lead fitted with an alligator clip at one end and a needle probe at the other. Jumpering across certain components or circuit sections is often the quickest means of isolating problems. For example: A transistor usually may be safely shorted from emitter to collector to represent a saturated condition, or from emitter to base to represent the cutoff state. It must be emphasized, however, that this cannot be done indiscriminately. It depends on the particular circuit and the components used. Shorting components in high-power output circuits is usually asking for trouble. Also, transistors and IC's can be damaged by applying voltages which exceed their maximum rating.

One of the most versatile and, surprisingly, seldom-used service aids is simply a jumper lead containing a series capacitor. A capacitor-resistor substitution box will, of course, serve the same purpose with even greater flexibility. However, a jumper lead with the capacitor built in offers the advantages of compactness, simplicity of use, and economy. Gadgets such as substitution boxes also have a habit of disappearing when you need them most.

Three jumper leads, constructed as shown in Fig. 1-1, will cover most troubleshooting applications. The uses of such a capacitor test lead are as follows:

1. *Bypassing.* Suspected emitter bypass or B+ decoupling capacitors can be easily checked by

59

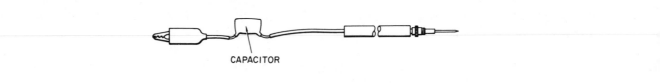

Fig. 1-1. Capacitor-isolated jumper lead.

jumpering. Poor filtering of agc networks can also be quickly located by this procedure.

2. *Signal injection.* Appropriate signals can be picked off at various points within a receiver and injected into other areas for test purposes. For example: Low-voltage 60-Hz ac can be injected into audio, video, i-f, and other stages as a test signal.

3. *Signal jumpering or criss-crossing.* Such a procedure is applicable to stereo audio amplifiers. Signals from one channel serve as useful test signals for injecting into suspected stages of the defective channel. This can be an extremely effective means of quickly isolating distortion problems occurring in one channel only.

4. *Signal grounding.* This is perhaps one of the most useful, and yet seldom-applied, functions of such a probe. Intermittent noises such as "popping" or "static-like" sounds due to intermittently shorted or open components or connections can be among the most frustrating problems to isolate. All technicians have spent many hours vainly tapping and flexing circuit boards and chassis, attempting to isolate a stubborn intermittent short or open connection. A more simple and effective approach is to isolate the intermittent condition to one circuit and often to one or two components. This can be accomplished in the majority of cases through use of a capacitor-isolated probe. The ac signal can be grounded through such a probe without disturbing the dc operating conditions. Working successively from stage to stage, the offending area often can be quickly isolated. As a simple example: Grounding the collector of a particular transistor through the capacitor probe eliminates the intermittent noise, while grounding the base does not. Obviously, the disturbance is being generated within the transistor itself. Incidentally, this occurs more frequently than generally realized. Transistors can generate intermittent popping and frying noises.

5. *Output meter.* A suitable capacitor, about .1 μF, connected in series with any ac meter converts the meter to an output meter. An output meter is simply one which reads the ac component of of a signal while blocking the dc present. An output meter is useful for signal tracing and in some circuits can serve as a substitute for an oscilloscope.

VTVMs AND VOMs

Contrary to popular opinion, a vom generally is more versatile for servicing solid-state equipment than a vtvm. The loading effect of the vom is not usually detrimental, since most solid-state circuits are of relatively low impedance. The base circuit can sometimes have sufficiently high impedance to introduce error. However, if the loading effect is anticipated, the slight error should not handicap the technician. The major advantage of the vom, when servicing solid-state equipment, is its current-measuring capabilities. A means of measuring current is at times mandatory, particularly when servicing battery-operated equipment.

Most voms employ two different voltages, one for high-resistance measurements and the other for low-resistance measurements. These voltages can be a convenient power source for biasing purposes, such as, turning a transistor on or off or clamping an agc line. Some voms also have a capacitor internally connected in series with the ac function, providing an output meter feature. Regardless of meter type, the technician must be familiar with the polarity and the value of the voltage at the test leads when using the ohms function. Many technicians condemn the use of the R×1 scale when checking transistors or transistor circuits. This is unfortunate since the R×1 scale is the most useful when checking many solid-state circuits. Any vom or vtvm producing a voltage of 1.5 volts or less at the test leads on the R×1 scale is safe to use for most testing purposes. In-circuit resistance measurements of transistors in direct-coupled amplifiers and output stages often can be successfully accomplished only on the R×1 scale. The low parallel impedances involved in such circuits make readings on the R×100 scale practically meaningless.

OSCILLOSCOPES

The "scope" is one of the most versatile pieces of equipment in the shop. Many problems cannot be logically solved without it. As previously indicated, outside service technicians often must develop techniques which, on the whole, are not equipment oriented. Conversely, the bench technician would, in most cases, be doing himself a favor by making signal tracing with an oscilloscope his primary service method.

The oscilloscope should be in good repair and calibrated to allow relatively accurate peak-to-peak voltage measurements. At least three probes should be available: a direct probe, a low-capacity probe, and a demodulator probe. Some shops build up an adaptor demodulator box which connects between the vertical input of the scope and the direct probe. An advantage to this method is that the detector may be constructed in the form of a doubler or even quadrupler circuit as shown in Fig. 1-2. This can be useful when signal tracing the low-level signals found in the first i-f amplifier stages, particularly if the scope lacks vertical sensitivity.

Recently, several low-cost triggered-sweep oscilloscopes have become available. The day is rapidly approaching when the triggered-sweep scope

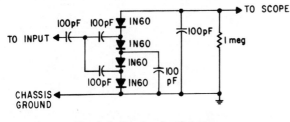

Fig. 1-2. Quadrupler detector.

will be considered an indispensable piece of shop equipment. Certain useful air-signal information cannot be observed with a conventional oscilloscope. The vertical interval test (VIT) signal is an example of such an air signal. The trend is toward using the VIT signal to an even greater extent, particularly in the area of automatic color control. The ability to view these signals enables the technician to quickly evaluate the response and alignment of the i-f amplifiers and the performance of the video amplifiers. Frequency measurements and peak-to-peak voltage measurements are also much quicker and easier with a triggered-sweep scope.

SIGNAL GENERATORS

A great variety of signal generators are on the market, some of which are combination types and others highly specialized. The trend today is toward individual crystal-controlled marker oscillators, particularly for color-tv i-f alignment. Regardless of the equipment selected, an efficient operation should have at least the following:

1. Sweep and marker generators adequate for color i-f amplifier and trap alignment.
2. Sweep and marker generators adequate for chroma alignment.
3. Sweep and marker generators for fm radio alignment.
4. A generator for a-m radio alignment.
5. A stereo multiplex generator.
6. A sine-wave oscillator. Some audio generators are available that operate at frequency ranges which are useful for servicing tv remote-control units (usually in the 30- to 40-kHz range).

CHAPTER 2

Power Supplies

Power supplies for solid-state circuitry are similar to those used in vacuum-tube circuits. However, the voltages are lower while currents are generally higher. The voltage regulation must also be better, as a general rule, to prevent shifting of bias and operating points. The full-wave bridge rectifier in Fig. 2-1 is the most-common power supply found in solid-state equipment. Full-wave bridge rectification allows the use of smaller and more economical power transformers. Full-wave rectification also reduces the size of filter capacitors since the output is 120 Hz.

The effective filtering action of such a rectifier and filter circuit is usually expressed as a percentage of ripple compared to the voltage across the load, and is approximately:

$$\text{percentage of ripple} = \frac{200,000}{RC}$$

where,
  R is the load resistance in ohms,
  C is the capacity in microfarads.

Generally, the value of the input capacitor is such that the ripple is about 5 percent of the load voltage. Further reduction in ripple content is accomplished by additional stages of RC filtering or electronic regulation.

A most-important and much-overlooked function of the power supply is that of decoupling. Technicians often tend to regard the power supply as merely a source of dc potentials, and forget the ac considerations. A well-designed power supply must appear as a low impedance at the frequencies involved. Since many circuits are often operated from the same source, the power supply must be the primary decoupling factor between the various circuits. Television power supplies should present a low impedance to both very low frequencies (tens of hertz) and very high frequencies (millions of hertz).

ELECTRONIC FILTERS AND REGULATORS

Purpose: Provide a relatively constant and ripple-free voltage over varying load and line fluctuations.
Signals In: 60-Hz (half wave) or 120-Hz (full wave) dc pulses.
Signal Out: Relatively ripple-free dc.

Circuit Description

Many solid-state circuits require a regulated power supply to prevent voltage fluctuations from changing the operating point. The most common circuits are the series regulator (Fig. 2-2A), the zener diode (Fig. 2-2B), or combinations of the two (Fig. 2-2C). Occasionally, a series regulator direct-coupled to an additional amplifier stage in a feedback configuration is used for more critical applications. All of these circuits are variously termed: *active filter, electronic filter, dynamic filter,* etc.

The simplest type of regulator and filter circuit, using a zener diode, is shown in Fig. 2-2B. Zener diodes are available in a wide choice of voltages ranging from about 3 volts to 100 volts. A zener diode behaves as any other silicon diode in the forward direction. However, in the reverse direction, insignificant conduction occurs until the characteristic zener voltage is reached. Conduction then occurs suddenly and a further voltage increase reduces the resistance of the diode, resulting in an essentially constant voltage across the device. The actual voltage increase for a typical zener diode is about 0.01 volt per milliampere of current increase. A zener diode acts much like a

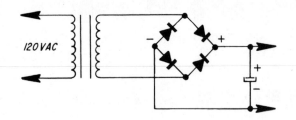

Fig. 2-1. Full-wave bridge rectifier.

battery in the circuit and as such appears as a low impedance to ac. For this reason, a zener regulator reduces power-supply ripple by a factor of about 10. Zener diodes have the disadvantage of not being economically available with large wattage ratings. The typical rating is 1 watt.

A zener diode may be direct-coupled to a transistor, as shown in Fig. 2-2C, to increase its power-handling capability. The base voltage of the series regulator is fixed by the zener voltage. The emitter voltage will follow the base voltage minus the 0.6-volt diode drop. The effective filtering action of the zener diode is multiplied by the current gain

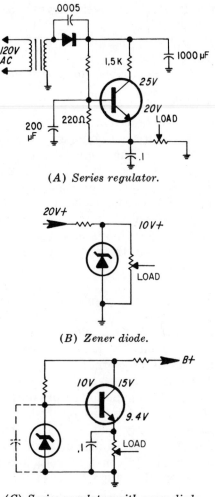

(A) Series regulator.

(B) Zener diode.

(C) Series regulator with zener diode.

Fig. 2-2. Typical regulator circuits.

of the transistor. If additional filtering is required, a capacitor can be connected from base to ground. The small amount of base current is easily filtered, resulting in a large ripple-free current at the emitter. The effectiveness of such a circuit is approximately equal to beta times the capacitance in microfarads. A beta of 100 and a capacitance of 100 microfarads would, therefore, be equivalent to a 10,000-microfarad capacitor.

A form of series regulator and filter can also be constructed without a zener diode, as shown in Fig. 2-2A. Such a circuit is responsive to load changes but does not regulate changes in line voltage. An increase in load demand causes an increase in bias current, reducing the emitter-to-collector resistance of the transistor. The transistor becomes a dynamic resistance in series with the load, maintaining a relatively constant output voltage. The filtering action is about beta times 200 μF. The 0.1 μF capacitor provides high-frequency bypassing. Large electrolytic capacitors have considerable inductive reactance and do not provide a sufficiently low impedance to the higher frequencies. Small capacitors with low inductive reactance are frequently found in such power supplies, particularly in tv receivers where decoupling high frequencies becomes an important factor.

The .0005-μF capacitor across the rectifier diode is to minimize so-called "silicon radiation." This phenomenon occurs because a silicon diode is such a good rectifier. When the ac sine wave reverse biases the diode, it cuts off very abruptly, accounting for the sawtooth waveforms shown in Fig. 2-3. The abrupt cutoff causes rapid current changes which shock associated inductive circuitry into momentary oscillation.

The result of this oscillation is high-frequency harmonics modulated at a 60-Hz rate. In radio circuitry, silicon radiation may become evident as a 60- or 120-Hz hum. In tv circuitry, silicon radiation appears as a dark or light horizontal band about three inches in height across the face of the crt. Silicon radiation usually becomes apparent only under weak-signal conditions. The capacitor across the diode should be considered a tuning capacitor and not a suppression or swamp-

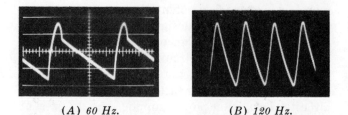

(A) 60 Hz.   *(B) 120 Hz.*

Fig. 2-3. Sawtooth waveforms with scope set at 30-Hz rate.

ing device. For this reason, a smaller value capacitor can often eliminate silicon radiation, whereas a larger value may intensify the problem.

Test Equipment

1. *A jumper wire.* Many useful tests can be performed simply by shorting certain components as described under service hints.
2. *A vom or vtvm.* A vtvm has no particular advantages over a vom when making voltage or resistance measurements in power-supply circuits, since loading is usually not a factor. A vom can, in fact, be advantageous since some of them feature an output-meter function useful for making ripple measurements. (This function consists of a blocking capacitor in series with the ac meter, allowing ac measurements exclusive of the dc component.)
3. *An oscilloscope.* A surprising fact to many technicians is that the oscilloscope is one of the most useful pieces of test equipment for troubleshooting the power supply. Certain problems are not readily observed by any other means. A common example is oscillation in high-gain amplifiers, caused by inadequate bypassing or decoupling in the power supply. All dc voltages can be normal and yet oscillations develop which can only be observed by viewing the B+ power-supply waveforms with an oscilloscope. Ripple is also easily measured and the difference between 60- and 120-Hz hum can be observed.

Service Hints and Procedures

1. A shorted (collector to emitter) or saturated series regulator transistor does not result in an inoperative set, since the load is in series with the transistor. However, the filtering action will be lost and hum will usually result. Television receivers will usually develop a hum bar across the crt face.
2. If a shorted regulator transistor is suspected, merely short the base to the emitter. No change would indicate that the transistor is indeed shorted.
3. If the set is inoperative and the filter circuit is suspected, momentarily short the emitter to collector. Restored operation indicates an open emitter-collector circuit, a shorted emitter-base diode, or the transistor is cut off because of improper bias.
4. One hum bar across the face of the crt indicates the problem is in the half-wave rectifier supply. Two hum bars indicate a problem in the full-wave rectifier supply. This becomes useful since some receivers have multiple power supplies, employing full-wave and half-wave rectifiers.
5. Motorboating or low-frequency oscillation in audio circuits is almost always caused by faulty electrolytic filter capacitors or defective decoupling capacitors on the B+ line.
6. A good habit to develop is to check B+ lines with an oscilloscope for the presence of signal. The B+ lines should be quite free of any signals. Signals present on B+ lines can cause feedback between circuits, which has resulted in all sorts of problems.
7. Many distortion problems in audio amplifiers can be traced to ultrasonic oscillations which develop due to inadequate decoupling in the power supply.
8. Care must be exercised when servicing circuits supplied from a series regulator. The entire load current flows through the transistor. A B+ short of even momentary duration will usually result in an open transistor.
9. Check the current drain before installing a new regulator transistor. A convenient means is to connect a dc ammeter across the open collector and emitter terminals. Start out on a high meter scale in the event that a B+ short does exist.
10. A suspected open or low-value filter capacitor can be checked by bridging a known good capacitor of comparable value across the component in question. Be careful to observe correct polarity and do not exceed voltage ratings. Also, always turn the set off before bridging filter capacitors in solid-state equipment. Transistors can be damaged by transient voltage spikes resulting from large charging currents.
    Note: Leakage between sections in multiple unit capacitors cannot be located by this method. Replacement of the entire suspected component is the best solution in these cases.
11. Most manufacturers include peak-to-peak ripple measurements on their schematics. Measurement of this ripple with an oscilloscope quickly indicates the condition of the power supply. When a particular circuit is suspected of drawing too much current, the ripple content of its B+ supply can be measured to verify this suspicion. The ripple content is directly proportional to the current drawn, as evidenced by the formula given at the beginning of this chapter.
12. When checking rectifiers in a bridge circuit, disconnect either end of each rectifier and

measure the front-to-back resistance ratio across each rectifier individually.

13. High-wattage resistors are notorious for changing values over a period of time. This is particularly true of wirewound resistors which tend to increase in resistance with age. Carefully check resistance values in cases of low B+ voltages.

14. When replacing high-wattage resistors, take note of their mechanical positioning. High-wattage resistors are often mounted in contact with the chassis or are placed in such a manner as to provide an air-chimney effect. Positioning in some other manner can cause premature failure due to heat fatigue.

15. Regulator transistors are usually mounted on a heat sink. When a replacement is made, be sure to mount the transistor mechanically secure and use heat-conducting greases where required. Be sure to install the proper insula-tors between the transistor and heat sink when required.

16. Before replacing open fuses, check the current drain by connecting an ammeter across the open fuse. Use the high-current range of the meter. Occasionally, fuses open for no apparent reason or because of line-voltage transients.

17. Stubborn cases of silicon radiation can often be solved by trying different values of capacitance across the silicon rectifiers. Try smaller-value capacitors as well as higher values. Rectifier and capacitor lead length is extremely important. Keep the leads as short as possible and close to the chassis. Lead length and placement of ac line chokes is also important. Lead dress of the conductors from the external antenna terminals to the tuner can be critical and should be kept as far from the power supply as possible.

REVIEW

Questions

Q1. Suppose the two biasing resistors in Fig. 2-2A have been transposed accidentally. The power supply is used in a tv receiver. What will be the probable effect on receiver performance?

Q2. Examine the schematic in Fig. 2-4. What will be the regulated output voltage to the load?

Q3. What will the output voltage to the load be if the 200-μF capacitor in Fig. 2-2A shorts?

Q4. Will the output voltage in Fig. 2-2A change if the emitter-base diode shorts?

Q5. List the three changes that will occur if a series-regulator transistor shorts from emitter to collector.

Answers

A1. A single hum bar will become apparent across the crt face. Reversing the biasing resistors will cause the transistor to saturate and the filtering action will be lost.

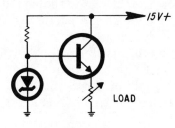

SILICON TRANSISTOR.
10 VOLT ZENER DIODE.

Fig. 2-4. Determine the ouput voltage to the load.

A2. The output voltage will be about zero. The zener diode is installed backwards, causing the diode to be forward biased. Therefore, about 0.6 volt will appear across the diode. The 0.6-volt drop across the emitter-base diode leaves an output of about zero volts.

A3. The output voltage will be about zero. The transistor cuts off, opening the B+ path to the load.

A4. Yes. The voltage will drop considerably because the transistor cuts off. The only path from load to B+ is now through the 1.5k biasing resistor.

A5. (a) The hum or ripple level will increase because the filtering action is lost.
(b) The voltage to the load will increase by an amount equal to the normal emitter-to-collector drop.
(c) The voltage to the load will fluctuate since regulation is lost.

CHAPTER 3

Tuners

Solid-state tuners very closely resemble tube tuners as far as circuit configuration is concerned. The major areas of difference are the bias circuitry and the means by which agc action is produced. The tuner is often a self-contained unit, and as such, is often considered a replaceable component. This has been true concerning the majority of uhf and vhf tuners. Most present-day solid-state tuners do not contain plug-in devices and, consequently, there is even a greater trend toward replacement of the entire unit. Most solid-state tuner failures are either mechanical in nature (dirty contacts, etc.) or due to failure of a solid-state device. The decision as to whether a tuner should be repaired or replaced depends primarily on the particular tuner involved. Transistor replacement in some tuners is economically feasible and in others it is a losing, if not impossible, proposition.

Tuners used in radio circuitry are usually an integral part of the receiver and, as such, must be repaired. Fortunately, the transistors are usually more accessible than those in tv tuners. In some radio receivers, fm tuners may also be considered replaceable components as in tv receivers. Most fm tuners are, however, more accessible and repairable than their tv counterparts.

Recently, many manufacturers have introduced electronic or *varactor* tuning in both radio and tv receivers. The objective of such systems is the mechanical aspects of station or channel selection. A varactor diode replaces the variable tuning capacitor in radio tuners and the switched inductances in tv tuners. Varactor diodes are available in a variety of capacitance ranges designed to operate cross the a-m, fm, vhf, and uhf bands. Basically, the varactor diode utilizes the variable width of its depletion region as a variable-width dielectric. Therefore, the capacity of the diode is proportional to the reverse-bias voltage. That is, as the reverse-bias voltage is increased, the capacity decreases. The use of variable-capacity diodes is by no means a recent development. The afc systems in fm radio and tv receivers have used such diodes for years. However, only recently have they been economically available in a wide range of values.

A-M TUNERS

Purpose: Selectively amplify the rf signal from the desired station and convert it to a center i-f frequency of 455 kHz.

Signal In: A broad spectrum of rf energy.

Signal Out: A 455-kHz carrier, amplitude modulated by the audio information of the station to which the receiver is tuned.

Circuit Description

The simplest form of a-m tuner is as shown in Fig. 3-1A. A single transistor performs the three functions of rf amplifier, oscillator, and mixer. Converter circuits such as this one are widely used in low-cost a-m radios. Basically, the circuit is a common-base oscillator with the rf signal applied in series with the base ground return. The stage is operated at maximum gain since no avc (automatic volume control) voltage is applied. Automatic volume control is achieved under most signal conditions by varying the gain of the first i-f amplifier stage. The audio-detector diode, D2, produces a positive rectified voltage which is filtered by C12 and applied as a reverse avc voltage to the base of Q2. The gain of Q2 is reduced as signal strength increases. As the conduction of Q2 decreases with increasing signal strength, the volt-

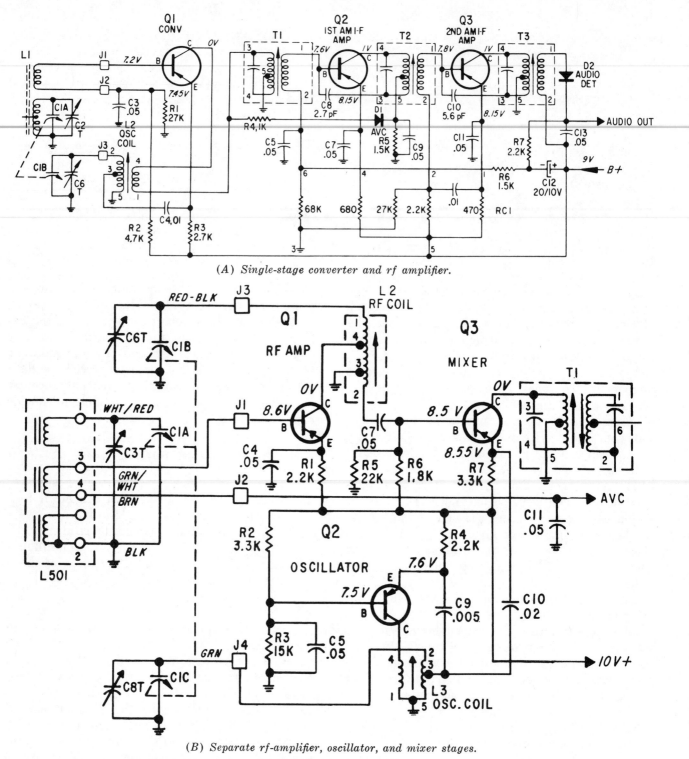

(A) Single-stage converter and rf amplifier.

(B) Separate rf-amplifier, oscillator, and mixer stages.

Fig. 3-1. Typical a-m tuners.

age developed across R5 also diminishes. D1 is normally reverse biased. Under very strong signal conditions, the voltage developed across R5 becomes sufficiently small to forward bias D1. The signal at the collector of Q1 now has a relatively low-impedance path to ground through C9, reducing the signal applied to Q2.

Test Equipment

1. Jumper lead containing about a .1-μF capacitor as shown in Fig. 1-1.
2. A vtvm or vom.
3. An oscilloscope.
4. Modulated a-m signal generator covering fre-

70

quencies of about 400 to 1700 kHz. Such a generator is necessary for rf and i-f alignment and also becomes a valuable troubleshooting aid.

Service Hints and Procedures

1. The most common malfunction to occur in a-m tuners is oscillator failure. The second most common failure is probably in the rf amplifier.

2. Transistors can be quickly checked by dc voltage measurements or front-to-back resistance ratio measurements. The most common cause of amplifier or oscillator malfunction is the transistor itself. The second most common cause of oscillator failure in a-m tuners is the oscillator coil.

3. A leaky transistor or a transistor with a low beta will also cause oscillator failure. Generally, such a condition can be found by alternately heating and cooling the device. Such a defective transistor will often oscillate at one temperature and not at another. A convenient means of applying heat is with a soldering gun or iron held in close proximity to the suspected device. Aeresol spray cans containing freon are available for cooling purposes.

4. Oscillator coils sometimes fail for no apparent reason and cannot be checked by ordinary means, unless the coil is open or is a dead short. Occasionally, the tuning range of a coil becomes impaired, resulting in poor sensitivity even though oscillation does occur. Trial replacement of the part is usually the quickest procedure in such cases.

5. Open or low-value bypass and coupling capacitors can be quickly checked by bridging with a capacitor probe, as illustrated in Fig. 1-1. The capacitors in Fig. 3-1B which could be checked by this method are: C4, C5, C9, and C10. The value of capacitance used in such a probe is not particularly critical since the objective is to identify the defective component. In this case, use about a .1-μF capacitor.

6. The oscillator can be viewed directly with an oscilloscope. In circuits with a separate oscillator (Fig. 3-1B), the oscillator signal is typically injected into the emitter circuit of the mixer. Because of the low impedance of the circuit, the oscillator signal can be viewed at this point with a direct probe without experiencing loading problems. Oscillator amplitude is important and fairly critical since conversion gain in the mixer is proportional to the oscillator signal. The oscillator signal at the emitter of the mixer should be about .5-volt peak to peak in most circuits of this type.

7. The oscillator operates at 455 kHz above the rf carrier. The oscillator section of the variable tuning capacitor will, therefore, be the smallest section in the tuning gang.

8. The instructions of the manufacturer should be followed for alignment. As a general rule, the oscillator coil is adjusted for maximum sensitivity at the low end of the band, and the oscillator trimmer capacitor is adjusted for maximum sensitivity at the high end. Again, the rf coil is adjusted at the low end of the band, and the rf trimmer at the high end. The antenna trimmer is normally adjusted for maximum sensitivity near the center of the band, usually about 1050 kHz.

9. Motorboating or low-frequency oscillation in a-m receivers is usually caused by a defective avc filter capacitor. (C12 in Fig. 3-1A.)

REVIEW

Questions

Q1. What would the effect be if D1 in Fig. 3-1A were accidentally installed backwards?

Q2. Would L2 in Fig. 3-1B be adjusted with the tuning gang open or closed?

Q3. What effect would an open C5 in Fig. 3-1B have?

Q4. What is the purpose of C4 in Fig. 3-1B?

A1. The sensitivity would be reduced at low to medium signal levels and overload would occur at high signal levels. The on and off mode of the diode would be exactly the reverse of what is desired.

A2. Closed. As a general rule, coils are adjusted at the low-frequency end of the band which requires maximum capacity, or a closed gang.

A3. The receiver would be dead except for noise. The common-base oscillator can no longer function, since the base circuit is open as far as an ac signal is concerned.

A4. C4 bypasses the emitter resistor of the common-emitter rf amplifier, preventing degeneration and loss of gain in this stage. An open capacitor would lower the gain of this stage drastically.

FM TUNERS

Purpose: Selectively amplify the rf signal from the desired station and convert it to a center i-f frequency of 10.7 MHz.

Signal In: A broad band of rf energy.

Signal Out: A 10.7-MHz carrier, frequency modulated by the audio information of the station to which the receiver is tuned.

Circuit Description

Most fm tuner circuitry is essentially the same as for a-m tuners. One distinguishing feature of fm tuners is the addition of an afc (automatic frequency control) circuit. The afc circuit consists of a feedback loop which causes the oscillator to seek frequencies resulting in zero dc-voltage output from the detector circuit. A varactor diode is used as a voltage-variable capacitor which is part of the oscillator tank circuit. The capacity of the diode decreases as the reverse bias across it increases.

An fm tuner may have separate rf-amplifier, oscillator and mixer stages as shown in Fig. 3-2A or may consist of an rf-amplifier stage and a converter stage as shown in Fig. 3-2B. The oscillator or converter stage is almost invariably of the common-base configuration. Many fm tuners, in fact, also utilize common-base rf-amplifier and mixer stages. The common-base configuration is so popular because it has the highest frequency cutoff. In other words, the gain of the common-base circuit at high frequencies can be greater than the common-emitter configuration. A recent trend has also been the use of field effect transistors in the rf-amplifier stage.

The avc (or agc) action is accomplished in much the same manner as in a-m tuners. A separate diode is used for fm because of the difference between the a-m and fm detection systems. The second i-f output signal is sampled, rectified, filtered, and fed back as reverse agc to the rf amplifier. In Fig. 3-2A, the diode is connected to produce an increasing negative voltage as the signal increases, since the rf amplifier is an npn. The opposite polarity is produced in Fig. 3-2B to produce reverse agc for the pnp rf amplifier.

Test Equipment

1. Jumper lead containing about a 0.01-μF capacitor.
2. A vtvm or vom.
3. A modulated fm signal generator covering frequencies of about 88 MHz to 108 MHz. Alignment of the rf and oscillator circuits requires such a generator. However, front-end alignment can often be accomplished reasonably well with an air signal.

Service Hints and Procedures

1. As in a-m tuners, the most common malfunction in fm tuners is transistor failure. Transistors can be quickly checked by dc voltage measurements or front-to-back resistance ratio measurements. Most transistor failures are shorted or open elements and are easily located by either method.
2. Oscillator and rf-amplifier failures are the most common problems. Converse to a-m tuners, oscillator failure in fm tuners is hardly ever due to a faulty oscillator coil which has a relative few number of turns.
3. Leaky transistors or transistors with low beta can cause oscillator failure or poor performance in the rf-amplifier or mixer stage. Alternate heating and cooling of the suspected de-

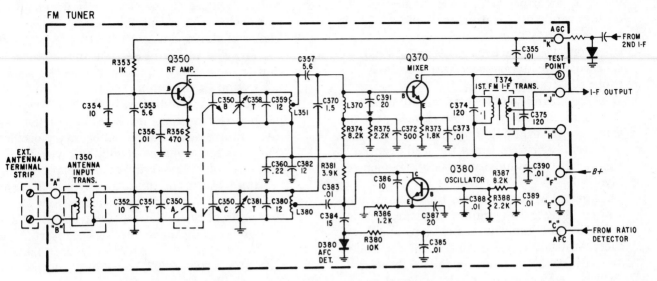

(A) Separate oscillator and mixer stages.

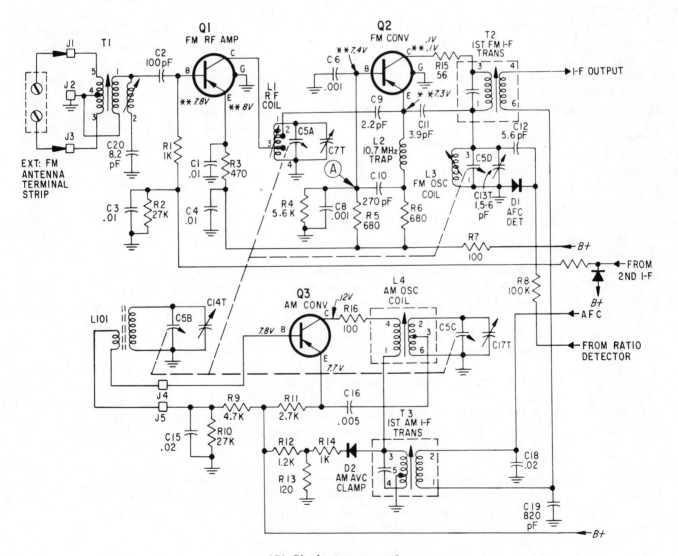

(B) Single-stage converter.

Fig. 3-2. Typical fm tuners.

vices, as described under a-m tuners, can be quite useful in isolating problems of this nature.

4. Open or low-value bypass capacitors can be checked by bridging them with the capacitor probe illustrated in Fig. 1-1.

5. When replacing components in fm tuners, take care to mechanically position the new part the same as the original. Transistor lead length is especially critical, since the leads can have significant inductances at the high frequencies involved.

6. The oscillator output of fm tuners cannot be viewed directly by ordinary means. An inoperative oscillator is easily detected, however. Adjust the treble control for maximum and adjust the volume control to about midposition. A very noticeable hiss should be apparent across the entire band if the oscillator is dead.

7. An inoperative mixer stage can also be detected in a similar manner. Adjust the treble and volume controls to maximum. Hiss will again be apparent across the entire band. A dead mixer can be distinguished from a dead oscillator by the volume control setting necessary to make the hiss audible. Audible hiss at about half volume indicates a dead oscillator, while hiss at full volume indicates a dead mixer.

8. An inoperative rf stage will usually result in extremely weak reception.

9. The oscillator operates 10.7 MHz above the rf carrier. The oscillator section of the variable tuning capacitor will, therefore, be the smallest section in the tuning gang.

10. A scope was not indicated in the list of recommended test equipment, since the frequencies involved are not viewable on ordinary oscilloscopes. A scope can be useful, however, to detect unwanted low-frequency oscillation. The B+ lines in an fm tuner are bypassed for a wide range of frequencies successively along the length of the conductor. In addition, chokes or damping resistors are often inserted in series with the B+ line. R7 in Fig. 3-2B is such a resistor. C360, C382, and C390 in Fig. 3-2A are examples of B+ bypass capacitors.

11. The manufacturer's instructions should be followed when performing alignment. As with a-m tuners, coils are usually adjusted at the low end of the band and trimmer capacitors are adjusted at the high end. If a modulated fm generator is not available, the oscillator can be set using two stations with known frequencies. One of the stations selected should be at the high end of the band and the other station should be at the low end of the band. The i-f alignment should always be performed before tuner adjustments are made.

REVIEW

Questions

Q1. Frequency modulation and amplitude modulation of an rf carrier are fundamentally different processes. Because of this, fm tuner design must be basically different from a-m tuner design. True or false?

Q2. In Fig. 3-2B, capacitors C6 and C8 are essentially in parallel. Why not just use a .002-μF capacitor from the base of Q2 to ground?

Q3. What is the purpose of C386 in Fig. 3-2A?

Q4. A receiver is inoperative on fm. A slight hiss is heard at maximum volume setting. What is the most likely problem?

Answers

A1. False. The type of modulation has nothing to do with tuner design. The function of the tuner is to selectively tune the rf carrier and reduce it to an i-f frequency. Both a-m and fm tuners are functionally identical and differ only in the frequencies processed.

A2. The two capacitors are not in parallel for the purpose of doubling the capacity. The conductor or printed-circuit foil from the junction of C6 to C8 represents a significant inductance at the frequencies involved and, as such, requires bypassing at both ends.

A3. C386 is the feedback capacitor which sustains oscillation. The common-base configuration is a natural oscillator since the input and output are in phase. Any capacitive coupling between the collector and the emitter causes oscillation to occur.

A4. The mixer or converter stage is inoperative. The slight hiss indicates that the i-f amplifiers are operating. A much louder hiss would indicate that the mixer stage is operating and the oscillator is dead.

VHF TV TUNERS

Purpose: Selectively amplify the rf signal from the desired channel and convert it to a center i-f frequency (44 MHz in modern receivers).

Signals In: A broad band of rf energy.

Signals Out: (1) A 45.75-MHz carrier, amplitude modulated by the picture information of the station to which the receiver is tuned. (2) A 41.25-MHz carrier, frequency modulated by the audio information of the station to which the receiver is tuned.

Circuit Description

Transistor vhf tuners invariably have three stages consisting of separate rf-amplifier, oscillator, and mixer sections as shown in Fig. 3-3. The operation is essentially the same as for the a-m and fm tuners previously described, except that tuning is accomplished by switching inductances rather than changing capacity. The agc circuit may also be different from that generally used in a-m and fm radio tuners.

Transistor gain can be controlled by either of two methods. These methods are generally referred to as forward agc and reverse agc. Transistor gain is related to emitter current as discussed fully in Part I of this book, and in Part II under agc systems. When a transistor is operated with an emitter current that produces maximum gain, an increase or decrease in emitter current reduces the gain. Each type of agc system has certain advantages, although the trend is to forward agc because it tends to overload less rapidly. The tuner illustrated in Fig. 3-3 employs forward agc. The voltage applied to the base of the rf amplifier becomes increasingly positive at greater signal strengths, causing transistor conduction to increase and gain to decrease. A common practice in tv tuners is to delay the application of agc voltage, or the effect of the agc voltage, until a certain

signal level is received. Usually, gain reduction does not occur until about 500-microvolts of signal is attained. This practice assures a good signal-to-noise ratio at low signal levels.

Most independent service dealers tend to consider the tuner as a replaceable module and only repair obvious defects. A number of organizations exist which specialize in tuner repair and rebuilding. This approach is usually more economical for the independent technician than attempting to repair tuners himself.

Test Equipment

1. A vom or vtvm. A vom can, at times, have a dual purpose since it serves as a convenient source of dc voltage. Many voms have two separate internal battery supplies for the low and high resistance ranges. A Simpson Model 260, for example, supplies 1.5 volts on the R×1 and R×100 scales and 7.5 volts on the R×10,000 scale. If the bias supply is not available, these voltages provide a simple means of clamping the tuner agc voltages. (See Chapter 1 on test equipment.)

2. An oscilloscope with a demodulator probe.

3. Alignment of the rf circuits requires a sweep generator covering the vhf band of 54 to 216 MHz and a marker generator for 41.25 MHz and 45.75 MHz. The i-f alignment also usually begins at the tuner since the mixer output inductance can be considered as the first i-f transformer.

Service Hints and Procedures

1. Most vhf tuner problems are mechanical in nature; dirty switch contacts being by far the most common complaint. Many cleaning solutions and aerosol spray solvents for cleaning dirty and corroded switch contacts are on the market. Some tuners contain plastic parts which can be damaged by some cleaning solvents. Before using any cleaner, be certain that

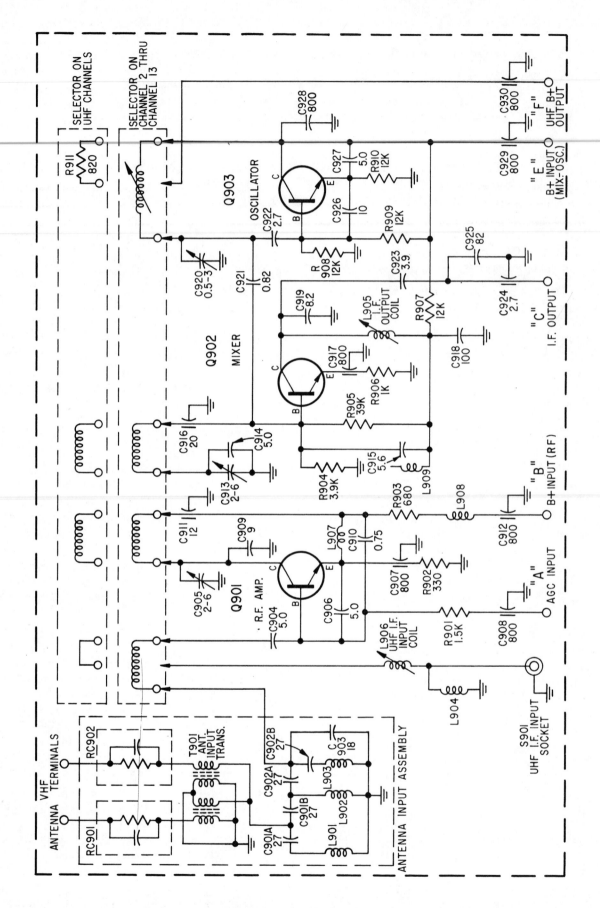

Fig. 3-3. Typical vhf tv tuner.

the label specifically states that the solvent will not harm plastics of any kind. The effectiveness of any spray is related to the mechanical accessibility of the switch contacts. In most cases, the tuner cover must be removed to do an adequate cleaning job.

2. A defective vhf tuner can produce many different symptoms. The most common are:
   A. No picture, milky raster.
   B. No picture, snowy raster.
   C. Overloaded picture (usually agc problem).
   D. Uhf reception but no vhf.
   E. Weak, snowy reception.

3. When a defective tuner is suspected, the first step is to check the B+ voltage at the tuner. A shorted transistor will usually drag this voltage down considerably. Transistor tuners are sometimes used in hybrid sets (sets using both tubes and transistors). In these cases, the B+ supply is usually derived from a divider network off a high B+ line. These divider resistors often open or change value. At this time, a quick check of the agc voltage may also be informative.

4. The second step is to clamp the agc voltage at approximately the voltage shown on the schematic. An adjustable agc bias supply is best for this purpose. However, an ohmmeter will usually suffice for test purposes, as described previously. If proper reception is restored under these conditions, the problem is related to the agc system and not the tuner.

5. Problems in vhf tuners are usually a result of transistor malfunction. The particular defective stage can usually be determined by simple troubleshooting methods. Whether the defective component can be replaced or not depends on the mechanical structure of the particular tuner and the skill and patience of the technician. In any event, when replacing transistors in tv tuners, particular attention must be paid to mechanical positioning and lead length.

6. In general, each stage produces different symptoms when failure occurs. The following assumes that only one stage at a time is inoperative and the other two stages are functioning normally.
   A. Inoperative rf amplifier—Picture extremely weak or nonexistent, depending on signal strength. Raster or picture very snowy. Sound will usually be normal if there is any picture at all. Normally, uhf reception will also suffer since the vhf rf amplifier becomes an i-f amplifier in the uhf position.
   B. Inoperative oscillator—No vhf reception; uhf reception may be normal. The raster is generally snowy. A loud rushing sound will be heard with the volume control set at about midposition.
   C. Inoperative mixer—No vhf or uhf reception. Raster is generally snow free but not totally devoid of noise. A hissing or rushing sound should be heard at the maximum volume setting. Assuming the audio circuits are operating properly, no hiss is usually indicative of a dead i-f amplifier stage.

7. Alternate heating and cooling of suspected devices as previously described is useful in detecting leaky transistors.

8. The i-f output from the tuner can be observed with an oscilloscope in most receivers. A detector probe must be used and the input cable to the i-f amplifiers disconnected. An alternate, and sometimes easier, method is to unplug the shielded output cable from the tuner and insert a dummy cable. The end of the dummy cable is then connected to the demodulation probe of the oscilloscope. Experience will soon educate the technician as to what constitutes a normal or inadequate signal level at this point.

UHF- TV TUNERS

Purpose: Selectively tune the rf signal from the desired channel and convert it to a center i-f frequency of 44 MHz.
Signals In: A broad band of rf energy.
Signals Out: (1) A 45.75-MHz carrier, amplitude modulated by the picture information of the station to which the receiver is tuned. (2) A 41.25-MHz carrier, frequency modulated by the audio information of the station to which the receiver is tuned.

Circuit Description

Most uhf tuners are mechanically quite simple and are very similar from one manufacturer to another. Fig. 3-4A depicts a tuner manufactured by Sarkes Tarzian, Inc. Notice the similarity to the uhf tuner manufactured by General Industries in Fig. 3-4B.

The uhf tuners, in common with vhf tuners, consist of three separate sections. However, only two of the uhf stages incorporate active devices. The rf or preselector stage is simply a rather broadly tuned circuit. The three stages, preselector, mixer, and oscillator are physically separated from each other by shielding. Energy is transferred between stages by ports cut in this

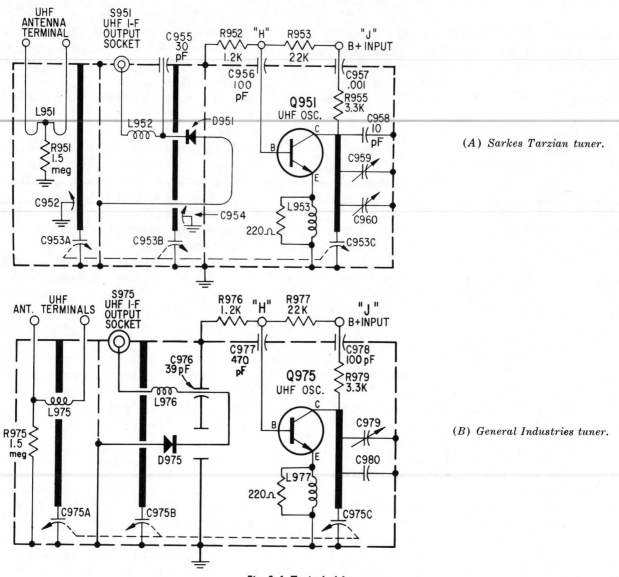

(A) Sarkes Tarzian tuner.

(B) General Industries tuner.

Fig. 3-4. Typical uhf tv tuners.

shielding. Each stage, in conjunction with these ports or windows, becomes a tuned cavity resonating across the band of frequencies desired (470 to 890 MHz). Notice that the General Industries tuner employs additional coupling between oscillator and mixer stages in the form of a half loop inserted from the mixer into the oscillator cavity.

Only two active devices are employed in uhf tuners; the mixer and oscillator. The mixer is simply a special high-frequency diode. The oscillator is a high-frequency transistor connected as a common-base oscillator. Oscillation is sustained by coupling between the collector resonator bar and the inductance in the emitter circuit. The oscillator operates 44 MHz above the selected rf carrier.

The i-f output of the uhf tuner is coupled to the rf amplifier of the vhf tuner. When the vhf tuner

is switched to uhf position, B+ is removed from the vhf oscillator, disabling this stage. The inductances tuning the rf-amplifier and mixer stages are switched to the 44-MHz range. These two stages now function as i-f amplifiers. The vhf antenna input circuit is switched from the vhf tuner and the i-f output from the uhf tuner switched in.

Test Equipment

1. A vtvm or vom.
2. Adjustment of uhf tuners is usually accomplished with air signals. Additional alignment equipment is not normally required.

Service Hints and Procedures

1. Check the B+ supply. The B+ supply in hybrid receivers is usually divided down from a higher B+ source.

2. Usually, uhf tuners are easier to repair than vhf tuners because of their mechanical simplicity. Due to the high frequencies involved, the inductors and other components are mechanically quite rugged. Failure is almost always due to the active devices; the oscillator transistor or the mixer diode.

3. Because of the high frequency of operation, extreme care must be exercised regarding the the lead length and mechanical positioning of the transistor and the diode.

4. Failure of either the oscillator or mixer in the uhf tuner will generally produce snow, and a loud rushing sound will be apparent at about one half of the maximum volume setting.

REVIEW

Questions

Q1. What is the function of the vhf oscillator in the conversion process when receiving uhf transmissions?

Q2. No reception is obtained on vhf channels; uhf is normal. What is the problem?

Q3. Both vhf and uhf are dead. There is no noise or snow in the raster. No hiss is apparent at full volume. What is the problem?

Q4. How can the rf amplifier and mixer stages in the vhf tuner function as i-f amplifiers?

Answers

A1. It has no function. B+ is removed and the oscillator is inoperative.

A2. The vhf oscillator is inoperative. Normal reception of uhf indicates that the rf-amplifier and mixer stages in the vhf tuner are operating as i-f amplifiers.

A3. Chances are that one of the i-f amplifiers is inoperative. Some hiss is usually apparent at full volume if all the i-f amplifiers are functioning.

A4. The inductances in the tuner circuits are switched to resonate at 44 MHz.

CHAPTER 4

Intermediate - Frequency and Video Amplifiers

Intermediate-frequency (i-f) amplifiers are quite similar regardless of applications. The i-f amplifiers for a-m, fm, and tv are of the same circuit arrangement, only the capacitive and inductive component values change because of the different operating frequencies. A number of different approaches to i-f amplifiers are employed by the various manufacturers. Each of several possible systems have advantages and shortcomings. The primary considerations are performance versus cost and the availability or development of suitable devices.

CIRCUIT CONSIDERATION

Most i-f amplifiers are of the common-emitter or common-source (FET) circuit configuration. Rarely, common-base amplifiers have been used even though the common-base configuration may have greater voltage gain at tv i-f frequencies. The chief objection to common-base amplifiers is their inherent instability (tendency to oscillate). Transistor replacement, for this reason, becomes critical and requires close matching of parameters.

Solid-state i-f amplifiers can be roughly divided into four basic categories as follows:

1. Common-emitter or common-source configuration, neutralized.
2. Common-emitter or common-source configuration, unneutralized.
3. Cascoded configuration.
4. Combination a-m and fm types which could be any of the above configurations, but are usually common-emitter, neutralized.

In general, i-f amplifiers are designed to amplify a specific band of frequencies of widely varying signal strengths. To accomplish this, particularly at high frequencies, several factors must be considered. These are: automatic gain control, neutralization, and Miller effect.

Automatic Gain Control

Automatic gain control (agc) of i-f amplifiers may be accomplished by either forward agc or reverse agc. Some receivers may use a combination of both techniques, applying forward agc to one stage and reverse agc to another. The recent trend, particularly if bipolar devices are used, has been to forward agc because of greater overload immunity. Several different mechanisms are involved regarding gain control of solid-state devices, the major factors being beta and impedance changes. Also, techniques of agc control applied to fm i-f stages may not be suitable for a-m i-f stages. Although fm i-f stages can be biased close to saturation which will cause clipping (limiting) to occur and provide gain reduction, clipping in a-m i-f stages would cause distortion. A complete description of agc operation can be found in the chapter on agc systems.

Neutralization

Bipolar transistors and field-effect transistors which have three elements are triode-equivalent devices and, as such, may require neutralization. High-gain common-emitter or common-source tuned amplifiers are unstable and tend to oscillate. This instability is a result of the capacitance and resistance existing between the input and output circuits. The most troublesome factor is the internal capacitance between the collector and base. This capacitance becomes quite significant, especially at higher frequencies. The amount of phase shift occurring across this internal RC network depends on the frequencies involved and the tuning of the output and input circuits. Phase shifts can occur which may cause either regeneration or degeneration. Regeneration causes oscillation if the gain is sufficient and degeneration results in loss of gain. The effect of this internal feedback signal can be cancelled or neutralized by feeding back an additional signal of equal amplitude but shifted 180° in phase. This can be accomplished in many different ways. Usually, a small-value capacitor is connected from the tuned output circuit back to the base in such a manner as to provide the 180° phase shift. The value of this capacitor ranges from about 1.5 pF to 10 pF and is dependent on the turns ratio of the i-f transformer. The most commonly used neutralization circuits are illustrated in Fig. 4-1. The circuit in Fig. 4-1A achieves the required degree of phase shift by connecting the feedback capacitor to the secondary of the i-f transformer. The circuit in Fig. 4-1B utilizes a tapped primary to provide feedback.

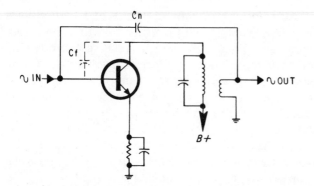

(A) *Feedback from secondary of the i-f transformer.*

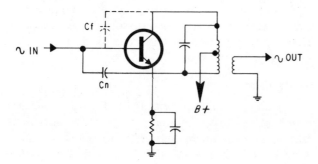

(B) *Feedback from tapped primary of the i-f transformer.*

Fig. 4-1. Neutralized i-f amplifiers.

Miller Effect

Not all i-f amplifiers are neutralized. Neutralization is required for relatively high-gain, high-impedance circuits. It is possible to trade gain for stability by loading the amplifier—reducing the input and output impedances. Not only is this possible, but is actually mandatory in high-frequency amplifiers because of a phenomenon known as the *Miller effect*. The Miller effect dictates that high-frequency voltage gain and high-impedance circuits are fundamentally incompatible.

The internal collector-to-base (drain-to-gate in the case of an FET) capacitance of most high-frequency transistors is about 1.5 pF. Considering stray external capacitance caused by leads, etc., this figure is more likely to be about 2 pF. Such a low capacity may not seem very significant until one realizes that this capacity is effectively multiplied by the voltage gain of the stage. That is, the capacity as seen by the input signal at the base is the collector-to-base capacity multiplied by the voltage gain of the stage. This is easily understood when one considers that the amplified signal voltage at the collector is essentially across the internal collector-to-base capacitance. This capacitance will draw signal current proportional to the signal voltage. The effect is exactly the same as if a much larger capacitor were connected from base to ground or B+. At high frequencies, this capacitance becomes a limiting factor in regard to high-frequency cutoff. This is best illustrated by example: Consider a television i-f amplifier designed to operate at a center frequency of 44 MHz. Assume a voltage gain of 100. The effective input capacity under these conditions would be about 200 pF (2 pF × 100). The capacitive reactance of 200 pF at 44 MHz is less than 20 ohms. The impedance of the signal source must, therefore, be less than 20 ohms or high-frequency cutoff will occur. (High-frequency cutoff is defined as a point where the current gain is 3 dB down from the maximum current gain). A higher signal-source impedance is feasible at these frequencies only by reducing the voltage gain of the stage through output loading or inverse feedback.

The example cited illustrates why high-frequency amplifiers are often not neutralized. The amplifier must be heavily loaded to achieve the necessary high-frequency response which precludes oscillation at the expense of gain. All high-frequency circuits must inherently be of relatively low impedance even though the active devices may have very high input impedances, as is the case with field-effect transistors. The Miller effect also explains why the common-base configuration has inherently the highest frequency response. Since the signal input is to the emitter and the base is essentially grounded, Miller effect is not a factor. Nevertheless, the input impedance is quite low, again requiring the typical low source impedance which is characteristic of high-frequency circuits.

Unneutralized Common-Emitter Configuration

A typical unneutralized tv i-f amplifier is illustrated in Fig. 4-2. The input is heavily loaded by the 100 pF capacitor at the input, preventing instability and oscillation. The output is single tuned and peak adjusted to the center of the i-f frequency. Bandwidth is obtained by the swamping resistor R_S connected across the tuned circuit. A forward agc system is employed, which causes transistor conduction to increase with increasing signal. The increased current results in lower beta and greater loading of the tuned circuit, which reduces gain.

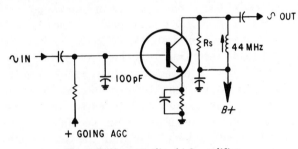

Fig. 4-2. Unneutralized i-f amplifier.

Cascoded Configuration

An alternate method of obtaining the required high-frequency response without sacrificing gain and without neutralization can be accomplished by connecting two transistors in a direct-coupled cascoded configuration.

The cascoded amplifier, as shown in simplified form in Fig. 4-3, is actually a common-emitter amplifier direct coupled to a common-base amplifier. This combination is ideally suited for a high-frequency amplifier, providing high gain at low noise. A higher input impedance is permissible without neutralization as a result of the high loading effect of the common-base input circuit on the common-emitter output circuit. The superior high-frequency characteristics of the common-base amplifier produce good voltage gain while providing an ideal no-signal point at the base for application of either reverse or forward agc.

Combination AM/FM I-F Amplifiers

A common practice in am/fm radio receivers is to use the same i-f stages for both a-m and fm

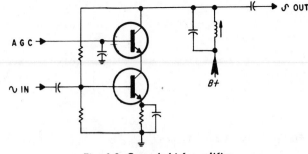

Fig. 4-3. Cascoded i-f amplifier.

amplification. Very little interaction occurs between the two-tuned circuits because of the widely separated frequencies involved. When operating at a-m frequencies, the inductive portion of the fm i-f transformer acts as a short circuit. (Refer to Fig. 4-4.) When operating at fm frequencies, the capacitance across the a-m i-f transformer acts as a short circuit. The 560-pF capacitors provide a ground return for the fm frequencies without significantly affecting the a-m frequencies.

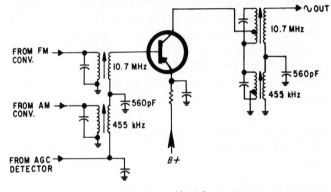

Fig. 4-4. An am/fm i-f stage.

A-M I-F AMPLIFIERS

Purpose: (1) Amplify a band of frequencies about 10-kHz wide, centered at 455 kHz. (2) Provide selectivity as a result of the frequency response of the i-f amplifiers. (3) Provide a relatively constant output over a range of varying signal strengths.

Signals In: (1) A 455-kHz carrier, amplitude modulated by the audio frequencies of the selected station. (2) A dc level proportional to signal strength, used to control the gain of one of the stages.

Signal Out: Amplified i-f signal, centered at 455 kHz and sufficiently narrow banded to provide good selectivity.

Circuit Description

The two stage a-m i-f amplifier, illustrated in Fig. 4-5, is the most common type found in home-

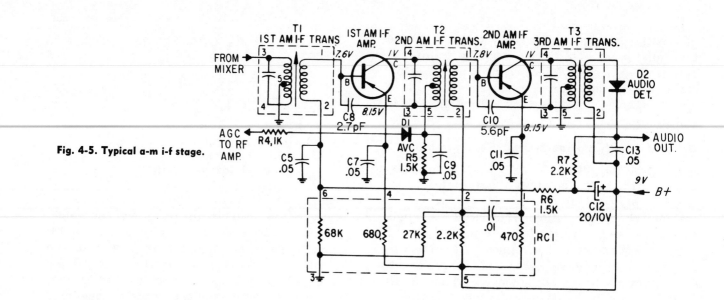

Fig. 4-5. Typical a-m i-f stage.

entertainment products. Because of the low frequencies involved, the Miller effect is not a factor and low-impedance loading is not required to achieve the necessary frequency response. The emitter resistors are bypassed with .05-μF capacitors. As in the case of high-gain amplifiers, neutralization is required in the form of feedback provided by the 2.7-pF and 5.6-pF capacitors. Reverse agc control is applied to the first i-f amplifier in the form of a dc voltage which becomes more positive with increasing signal strength. This voltage is derived by sampling a portion of the detected audio signal which is filtered by capacitor C12.

Test Equipment

1. Jumper lead containing a .1-μF capacitor as illustrated in Fig. 1-1.
2. A vom or vtvm.
3. An oscilloscope.
4. A signal generator providing a modulated 455-kHz output.

Service Hints and Procedures

1. Most a-m i-f amplifiers are high-gain stages. The output of the first a-m i-f stage is generally viewable with most service-type oscilloscopes. One of the quickest troubleshooting techniques is to simply signal trace the air signal through the i-f stages with an oscilloscope. The a-m i-f stages can be expected to produce a voltage gain of at least 10 and often much more, depending on design.
2. Dead i-f stages are characterized by lack of hiss at high volume settings. Dead stages are usually caused by transistor failure. Voltage measurements or a quick check of the front-to-back

resistance ratio of the transistor junctions will pinpoint the defective device in most cases.

3. Weak i-f stages may be caused by a defective emitter-bypass capacitor. Grounding each emitter with a capacitor probe, as shown in Fig. 1-1, will restore normal operation in these cases.
4. Weak a-m reception may be due to improper alignment or a defective i-f transformer. Inability to peak a transformer usually indicates that the transformer or the capacitor across it is defective. Alignment thus becomes a good troubleshooting technique and the signal generator becomes a convenient signal source for signal injection and tracing.
5. Some audio generators oscillate as high as 455 kHz and may be used as a signal source for troubleshooting or alignment.
6. Another convenient signal source is simply the 60-Hz ac from the power supply. Connect the capacitor probe to a low-voltage 60-Hz source, such as the power transformer, and inject it into the base of each i-f amplifier successively. A progressively louder hum should be heard, working from the output stage back to the mixer, if the stages are operating normally. A dead stage can be quickly isolated in this manner most of the time.
7. Motorboating or low-frequency oscillation is usually caused by inadequate B+ filtering or a defective agc filter capacitor. Bridging B+ and agc lines with about a 20-μF electrolytic capacitor will isolate such problems.
8. Noises such as popping or staticlike sounds due to intermittent components or connections can be among the most frustrating problems to isolate. The capacitor probe previously described can be most useful in tracking down

such intermittents. The ac signal can be grounded through such a probe without disturbing the dc operating conditions. Working successively from stage to stage, the offending intermittent can be isolated to the single stage and often to a single component.

REVIEW

Questions

Q1. What are the two functions of the detector diode in the output of the second i-f stage shown in Fig. 4-5?

Q2. Is the agc control of the i-f amplifier shown in Fig. 4-5 of the forward or reverse type?

Q3. The purpose of C5 in Fig. 4-5 is to load the input of the first i-f amplifier, swamping out the capacitance produced by the Miller effect. True or false?

Q4. Signal tracing with an oscilloscope through the a-m i-f stages requires the use of a detector probe. True or false?

Q5. What is the purpose of C13 in Fig. 4-5?

Answers

A1. Rectify (detect) the modulated rf signal to extract the audio signal. Develop a dc voltage for agc which becomes more positive with increasing signal strength.

A2. Since the dc is developed at the cathode of the detector diode, the voltage is positive going. The first amplifier is a pnp type. Application of a positive-going voltage causes less conduction. The agc control, as a result, is of the reverse type in which gain loss occurs through reduction of collector current.

A3. False. C5 bypasses the agc line, removing any rf frequencies which could cause oscillation.

A4. False. A detector probe could be used but is not necessary because of the relatively low frequencies involved. Most scopes can easily display the 455-kHz carrier with its audio-modulated envelopes (Fig. 4-6).

A5. C13 filters out the 455-kHz carrier from the detected audio signal.

FM I-F AMPLIFIERS

Purpose: (1) Amplify a band of frequencies about 200-kHz wide centered at 10.7 MHz. (2) Provide selectivity as a result of the frequency response of the i-f amplifiers. (3) Provide a relatively constant output over varying signal strengths.

Signals In: (1) A 10.7-MHz carrier, frequency modulated by the audio frequencies of the selected station. (2) A dc level proportional to signal strength, used to control the gain of one or more stages.

Signal Out: Amplified i-f signal, centered at 10.7 MHz and sufficiently narrow banded to provide good selectivity.

Circuit Description

As illustrated in Fig. 4-7, fm i-f amplifiers are generally of the common-emitter, neutralized configuration. Since the three i-f stages are identical, the degree of neutralization required for each

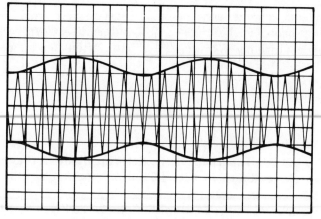

Fig. 4-6. Waveform in a-m i-f stage.

stage is the same and is provided by the 2.2-pF capacitors.

A limiter stage is employed even though limiting is not absolutely essential when a ratio detector is used as the demodulator. However, the use of a limiter stage in addition to the ratio detector provides even greater noise immunity.

The 10.7-MHz carrier is detected by D51 and filtered by C63, developing a dc voltage which becomes more positive at increasing levels of the

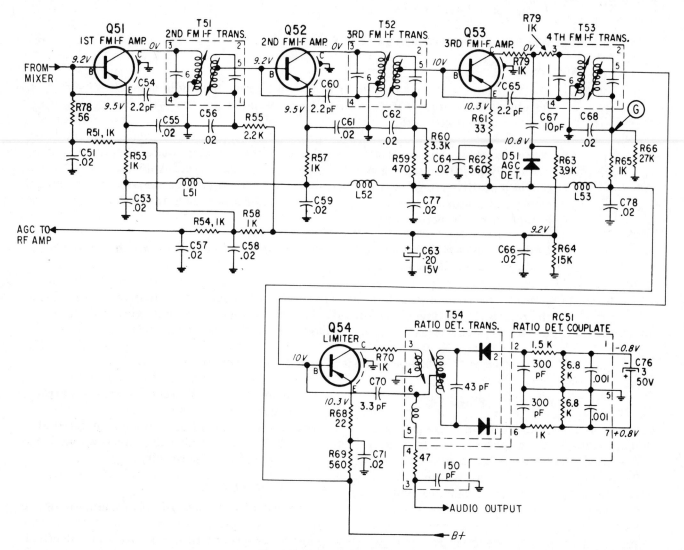

Fig. 4-7. Typical fm i-f amplifier.

86

10.7-MHz carrier. This positive voltage, applied to the base of the first and second i-f amplifiers, causes reverse agc action to occur—conduction decreases as the signal increases. Good signal decoupling in the agc line is essential, which is the purpose of the 1k resistors and 0.02-μF capacitors. Decoupling in the B+ line is also important to prevent oscillation. The rf chokes L51, L52, and L53 in conjunction with the 0.02-μF capacitors perform this function.

Test Equipment

1. A jumper lead containing a 0.1-μF capacitor as illustrated in Fig. 1-1.
2. A vom or vtvm. A vtvm is recommended for setting the ratio-detector transformer.
3. An oscilloscope with a demodulation probe.
4. A signal generator providing a sweep output at 10.7 MHz and markers at 10.7 MHz, 10.6 MHz, and 10.8 MHz.

Service Hints and Procedures

1. The fm i-f air signal can be easily traced through the i-f amplifiers with a detector probe and an oscilloscope. Since each stage is fully bypassed and neutralized, it can be expected to produce significant voltage gain (10× or more). The fm signal, before limiting, contains enough amplitude modulation to make signal tracing in this manner possible. This is particularly true if the tuning gang is adjusted slightly off station. The output from the limiter may also be observed in this manner when tuned to a station sufficiently weak as not to cause limiting action. A good indication that limiting action is indeed taking place is to view the limiter output using a detector probe. The recovered audio signal should be weak and about equal when properly tuned on all strong local stations. A significantly greater amount of recovered audio signal will be apparent on weak stations which cannot drive the stage into limiting. If all stations are producing a large amount of recovered amplitude modulation (as compared with the previous stage) from the limiter, and this varies considerably from station to station, indications are that the stage is not limiting. Since the design must be such that normal operation will produce limiting on local stations, we must assume that the receiver is defective (low gain). The low-gain condition may be isolated, if occurring in the i-f amplifiers, by signal tracing as just described. Weak reception may also be due to improper alignment, improper antenna, or defective tuner—particularly the rf amplifier.

2. An alternate method of signal tracing through the i-f stages is the audible method. A detector probe, or simply a detector diode such as a 1N60 connected in series with a 10k resistor, and a high-gain audio amplifier will allow audible tracing. Such a probe could even be constructed in the same manner as the capacitor probe illustrated in Fig. 1-1, and the audio portion of the receiver could be used as a test amplifier. The output of the first i-f amplifier should be just barely audible in most receivers.

3. Dead i-f stages are characterized by a lack of hiss with the volume and treble controls set high. Dead stages are usually caused by transistor failure. Voltage measurements or a quick check of the front-to-back ratio of the transistor junctions will isolate the defective device in most cases.

4. Weak or dead i-f stages can be caused by defective emitter bypass capacitors or oscillation may develop because of defective decoupling capacitors. Ground each emitter and the B+ and agc lines with the 0.1-μF capacitor probe shown in Fig. 1-1.

5. Alignment is a good method of troubleshooting weak i-f stages. Usually a defective stage will not align properly. Some manufacturers specify peak alignment as an alternate to sweep alignment. Most transistor i-f amplifiers are inherently of sufficient bandwidth and lend themselves well to peak alignment. One technique which can be used on many receivers employing stereo-multiplex circuitry, is to peak the i-f transformers for maximum recovered 19-kHz pilot signal when tuned to a stereo station. The 19-kHz signal can be observed with a scope or measured with a meter at the collector of the 19-kHz amplifier in the multiplex circuit. While this technique is not generally recommended, it usually works out so well that any performance degradation cannot be detected without special equipment.

6. Some low-cost, triggered-sweep oscilloscopes are presently on the market which are capable of viewing the 10.7-MHz i-f signal directly. Such instruments are invaluable for troubleshooting most electronic circuits. When signal tracing with a direct probe, it is often advantageous to connect about a 10k resistor in series with the scope.

7. Intermittent components or connections can be isolated by grounding the signal with a .1-μF capacitor probe.

8. Most transistors used as fm i-f amplifiers have

four leads, one of which is attached to the case. This lead is grounded to prevent 10.7-MHz radiation. In some cases, an external grounded shield is used for the same purpose. Be certain to reinstall these shields when replacing transistors which use them.

REVIEW

Questions

Refer to Fig. 4-7.

Q1. Why is a separate diode used as an agc detector instead of using the dc voltage developed by the ratio-detector circuit?

Q2. What is the purpose of C55 and C61 in the first and second i-f stages?

Q3. Why is a demodulator probe required when signal tracing through the fm i-f amplifiers?

Q4. All fm i-f amplifiers are specifically designed to reject any amplitude modulation and for this reason, fm reception is relatively noise free. True or false?

Q5. A simple diode detector can be used to signal trace the air signal through the fm i-f amplifiers, even though the modulation is in terms of frequency variations. True or false?

Answers

A1. The voltage developed across C76 would be proportional to the rf signal and could be used for agc control if it were not for the fact that limiting occurs in the previous stage. In some receivers employing four i-f stages, the agc voltage is developed from the emitter of the third i-f stage, eliminating the need for a separate diode.

A2. Emitter resistor bypass capacitors. The i-f amplifiers are neutralized stages capable of high gain without oscillation. The emitters are, consequently, fully bypassed to obtain this gain.

A3. Because of the high frequencies involved. Most service scopes will not respond to 10.7 MHz and, as such, the i-f carrier cannot be viewed directly. Demodulation enables viewing of the amplitude-modulated component of the i-f signal.

A4. False. The fm i-f amplifiers amplify a-m signals as well as fm signals. Noise immunity is provided by limiting either in the detector, the final i-f stage, or both.

A5. True. A significant a-m component exists, particularly when the receiver is slightly detuned.

TV VIDEO I-F AMPLIFIERS

Purpose: (1) Amplify a band of frequencies about 5-MHz wide and centered at approximately 44 MHz. (2) Provide selectivity and interference-free reception as a result of the frequency response of the i-f amplifiers and the various trapping circuits. (3) Provide a relatively constant output over varying signal strengths.

Signals In: A 45.75-MHz carrier, amplitude modulated by the video information of the selected channel. (2) A 41.25-MHz carrier, frequency modulated by the audio information of the selected channel. (3) A dc level proportional to signal strength which controls the gain of one or more stages.

Signals Out: Amplified i-f signal to the video detector centered at about 44 MHz and approximately 5-MHz wide.

Circuit Description

The i-f amplifier shown in Fig. 4-8A is unique in that it employs all dual-gate field-effect transistors. There has recently been a trend to the use of such devices as high-frequency amplifiers because of their inherent high-frequency gain combined with a high degree of stability. Each amplifier is actually a cascoded stage such as described at the beginning of this chapter. The transistor represented by gate 1 is connected as a fully bypassed common-source amplifier. The second transistor is of the common-gate configuration, evidenced by the fact that gate 2 is placed at signal ground by capacitors C319, C325, and C333. No neutralization is required and yet good stability and gain are realized because of the heavy loading of the common-source amplifier by the low input impedance of the common-gate amplifier. This particular i-f strip is incorporated in a color receiver, and as such, the interstage tuned circuits are stagger tuned, providing proper bandwidth and gain at the chroma and video frequencies.

Adjacent-channel sound and video trapping is accomplished in the input circuit. L312 and L313 are adjusted to resonate at 47.25 MHz, providing maximum attentuation of the adjacent-channel sound. L315 traps out 39.75 MHz, the adjacent-channel video. L314 is tuned to 44 MHz, the center i-f frequency. L310 and R311 are impedance-matching components for the i-f input cable. The 41.25-MHz sound carrier is attenuated by L332 in the output module. The 4.5-MHz sound carrier is detected by an additional diode, which is characteristic of color-tv i-f systems. Separate detection in this manner is the simplest means of minimizing the 920-kHz beat appearing in the picture. This beat develops between the 41.25-MHz and 42.17-MHz frequencies before detection. It develops between the 4.5-MHz and 3.58-MHz frequencies after detection.

A reverse agc system is used. Most of the agc is accomplished in the first i-f stage. A small amount of gain control is provided in the second i-f stage.

The third i-f stage is operated at maximum gain at all times, due to a fixed voltage of about 4 volts developed by the divider network and applied to G2 of Q330. This voltage is also applied to G2 of the first and second i-f stages, causing these stages to also operate at high gain during the absence of any applied agc voltage. As signal strength increases, a progressively more-negative voltage is applied to the gate of the first and second i-f stages, causing gain reduction. The voltage at gate 2 of the first i-f stage varies from about +4 volts under no-signal conditions to about −0.5 volt under very strong signal conditions, providing about 40 dB of gain reduction. The gain reduction of the second i-f stage with a strong signal is only 6 dB.

Fig. 4-8B depicts a bipolar-transistor i-f amplifier used in a black-and-white tv receiver. Several differences are apparent when compared to the i-f amplifier used in the color receiver. The overall bandwidth is neither as great nor as critical, allowing peak-tuned rather than stagger-tuned i-f transformers. Notice that each resonant interstage circuit is peak aligned to 44.15 MHz, the center i-f frequency. Sufficient bandwidth is provided by the swamping resistors R304, R314, and R323 which are connected across the tuned circuits. Trapping is also not as critical and only the adjacent-channel video is attenuated by L301. The 41.25-MHz sound carrier is attenuated at the input of the i-f strip rather than the output as was the case in the color receivers. This, again, is due to 920-kHz beat consideration.

Each stage is fully bypassed and yet neutralization is not required. This is possible because gain is traded for stability by loading methods. This is a comomn practice at the frequencies involved (44 MHz), since Miller effect dictates low circuit impedance anyway. The i-f stages are heavily loaded by the 100-pF capacitors from base to ground. These represent an impedance of less than 50 ohms at 44 MHz.

A forward agc system is employed in Fig. 4-8B. As the signal increases, a greater potential is applied to the base of the first i-f amplifier, causing increased conduction and reduced gain. Gain reduction occurs primarily because the transistor is being operated at an emitter current producing maximum high-frequency beta. Any increase of emitter current causes a rapid fall off in beta.

The tuned circuit views the transistor as a load as well as a source impedance. As the emitter-collector resistance decreases, the swamping effect across the tuned circuit causes increased bandwidth and gain decreases.

Test Equipment

1. A jumper lead, as illustrated in Fig. 1-1, containing a 0.01-μF capacitor.
2. A vom or vtvm. A vom meter is somewhat more versatile since it provides voltage sources suitable for clamping the agc system.
3. An oscilloscope with a demodulator probe or quadrupler circuit as illustrated in Fig. 1-2.
4. An adjustable bias supply for clamping the agc voltage.
5. A sweep generator and marker generator with marker frequencies at 39.75, 41.25, 41.67, 42.17, 42.67, 44.00, 45.75, and 47.25 MHz.

Service Hints and Procedures

1. Most i-f amplifier problems can display a great variety of different symptoms. In general, if an i-f problem is suspected, the first step is to determine whether the problem is directly a result of a defective i-f stage or is caused by improper agc action. Clamp the agc voltage at about the level indicated on the schematic. A variable dc-voltage source such as a bias supply is best for this purpose. If proper operation is restored at a particular bias voltage, the problem is most likely related to the agc circuit rather than the i-f amplifiers.
2. One of the best and simplest troubleshooting methods is to signal trace an air signal through the i-f amplifiers with an oscilloscope. The output of the first i-f amplifier is usually viewable with most oscilloscopes, particularly if the quadrupler detector illustrated in Fig. 1-2 is used. Each i-f stage should exhibit a significant voltage gain, generally 10 or more.
3. Low or no gain can be caused by defective emitter or source bypass capacitors. Bypass each emitter or source circuit with the capacitor probe if low gain is encountered. The bypass capacitor is defective if gain returns to normal.
4. Defective transistors are among the most common causes of i-f amplifier failures. Voltage measurements or front-to-back ratio measurement of bipolar transistors usually isolates the defective device.

5. In color receivers, the 3.58-MHz oscillator can sometimes be used as a signal source for injection into the i-f amplifiers.

    Disable the color killer or turn the color-killer control to the position which makes the circuit inoperative. Rotate the color or chroma control fully clockwise.

    Connect the capacitor probe to some low-impedance point in the 3.58-MHz oscillator circuit. (In most sets, the secondary of the 3.58-MHz transformer is okay.) With the probe end, inject the 3.58-MHz signal into the base or gate of each i-f amplifier successively. A color raster will result if the i-f stage is operating. The particular color will change from one i-f stage to another. The degree of color saturation and the particular color displayed depends on a variety of factors. The technician must gain experience with this method as applied to the particular set being serviced in order for it to be useful. With some practice, a dead or weak i-f stage can be readily detected by this technique.
6. Some i-f problems can be isolated only by alignment. The shop that is properly equipped to do alignment on a routine basis will seldom experience difficulty in solving i-f problems. Routine alignment checks also prevent borderline cases from leaving the shop and resulting in callbacks.
7. Do not underestimate the effect that shielding has on the performance of the i-f amplifiers. Removal of a shield surrounding an i-f transformer can detune the circuit sufficiently in some cases to cause the loss of color or deteriorate the picture a great deal. Shields over a group of tuned circuits can actually be the means of mutual coupling between the circuits. Removal of the shield in such cases may result in complete loss of picture. An example of this is the shield surrounding the output module in Fig. 4-8A.
8. A completely dead i-f stage will usually result in a noise-free raster and no sound. Certain failures in receivers employing direct coupling through the video amplifiers to the crt can cause loss of raster.

REVIEW

Questions

Q1. Why is the output to the sound circuits after the video detector in Fig. 4-8B and before the video detector in Fig. 4-8A?

Q2. If the i-f transformers in Fig. 4-8B are peak aligned, how is sufficient bandwidth obtained?

Q3. What effect will R314 in Fig. 4-8A have on the gain of the first i-f stage?

Q4. What effect will shorting gate 2 of the second i-f in Fig. 4-8A to ground have?

Q5. What is the function of C318, C323, and C338 in Fig. 4-8A?

Q6. Why are R316 and R300 separate 100-ohm resistors in Fig. 4-8A? Why not just use a single 220-ohm resistor?

Q7. Define forward and reverse agc.

Answers

A1. It is customary to extract the sound before the video detector in color receivers. The 41.25-MHz sound carrier can then be trapped out before video detection, minimizing the 4.5-MHz signal appearing after detection. The small remaining 4.5-MHz signal can now be readily trapped, preventing it from beating with the 3.58-MHz signal and producing a 920-kHz pattern in the picture. Black-and-white receivers often extract the 4.5-MHz sound carrier at the first video amplifier, utilizing the additional gain of this stage.

A2. The transformers are broad banded by connecting swamping resistors R304, R314, and R323 across the transformer primaries.

A3. It reduces the gain of this stage by employing negative or degenerative feedback. Any signal appearing on the base or gate of a grounded-base amplifier is of degenerative polarity.

A4. The gain will be reduced if operating under weak signal conditions and increased if operating under very strong signal conditions. Shorting the gate to ground places the gate at zero potential. Under very strong signal conditions, the agc voltage present at this point may be negative with respect to ground. Shorting the gate to ground would, therefore, cause a gain increase.

A5. These capacitors are actually part of the parallel tuned circuits which include the interstage i-f transformers L311, L321, and L331. The lower end of these inductors are at signal ground, which places the 15-pF capacitors essentially across them.

A6. A common practice in high-frequency circuits is to use resistors in lieu of jumper wires. A carbon resistor is essentially noninductive and, as such, dampens any tendency toward oscillation and radiation.

A7. Forward agc systems cause gain reduction by increasing the conduction of the active device. Forward agc is commonly used with bipolar devices. Reverse agc systems cause gain reduction by decreasing conduction of the active device. Such systems are generally employed when FETs or vacuum tubes are the active devices.

91

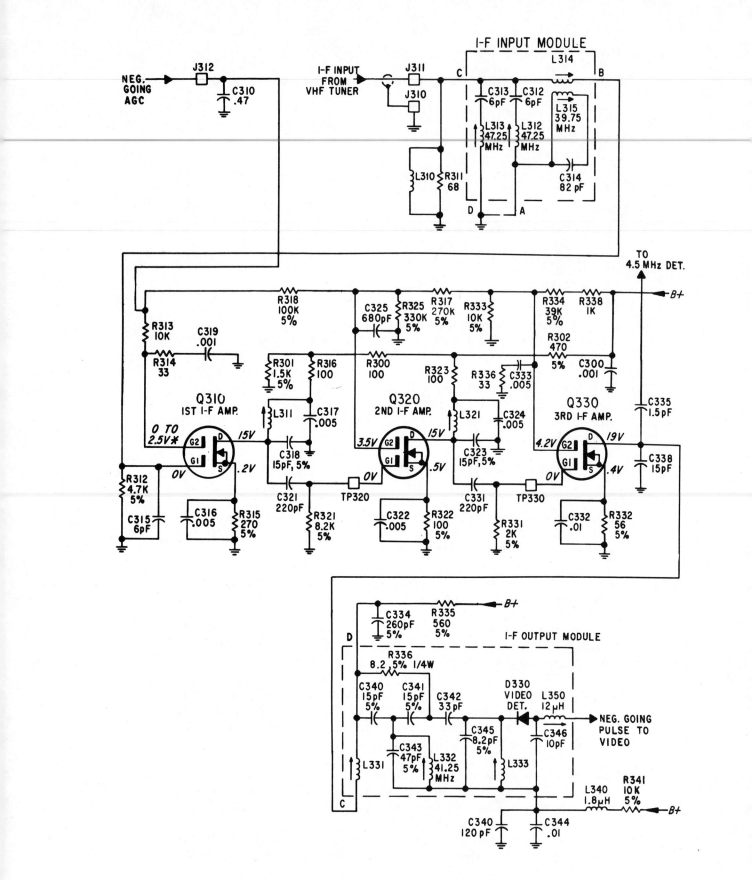

(A) An i-f amplifier using dual-gate FETs.

Fig. 4-8. Typical

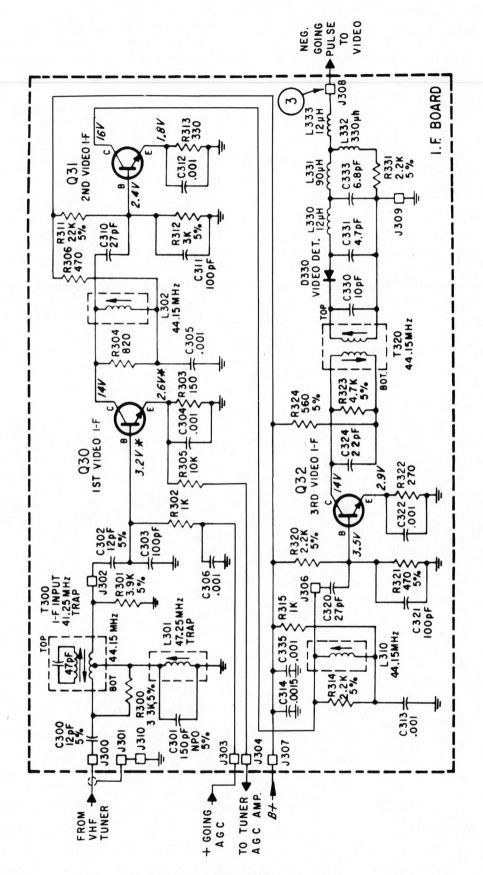

(B) An i-f amplifier using bipolar transistors.

tv i-f amplifiers.

TV SOUND I-F AMPLIFIERS

Purpose: (1) Amplify the 4.5-MHz sound i-f signal. (2) Most sound i-f amplifiers are designed to provide limiting action for better noise immunity.

Signal In: A 4.5-MHz carrier, frequency modulated by the audio portion of the selected channel.

Signal Out: Amplified 4.5-MHz carrier of constant amplitude.

Circuit Description

The circuit illustrated in Fig. 4-9 is a sound i-f amplifier used in a black-and-white tv receiver. The sound takeoff point is typically after the video detector in black-and-white receivers. The 4.5-MHz sound carrier is present at the takeoff point, eliminating the need of an additional detector diode.

The i-f transformer, T200, is rather sharply tuned since the bandwidth required for tv sound is only 25 kHz on either side of the carrier. T251 is the 4.5-MHz trap and sound takeoff transformer. The stage is fully bypassed and neutralized, producing high gain. The collector load resistors are of about the same value as the emitter resistor, causing limiting to occur when the stage is driven adequately.

Test Equipment

1. A vom or vtvm.
2. A jumper lead containing a 0.1-μF capacitor.
3. An oscilloscope with demodulator probe.

Service Hints and Procedures

1. The identical signal tracing and injection techniques may be employed as described for fm i-f amplifiers.

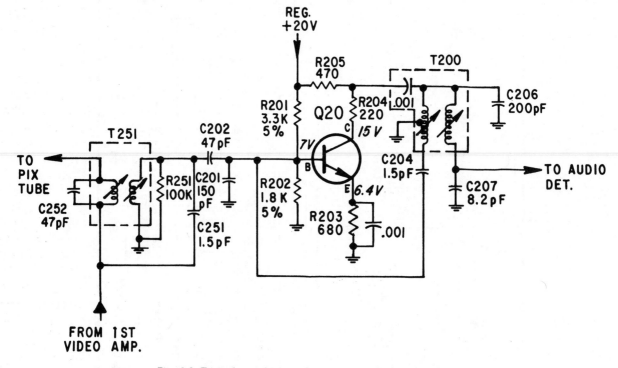

Fig. 4.9. Typical sound i-f amplifier used in black-and-white tv.

REVIEW

Questions

Q1. What is the purpose of T251 in Fig. 4-9 and what occurs if it is misadjusted?

Q2. What is the purpose of C204 in Fig. 4-9?

A1. T251 is the 4.5-MHz trap. If detuned, the 4.5-MHz sound carrier beats against the higher video frequencies, producing sound bars which vary with the audio information. Since T251 is also the sound takeoff transformer, the sound would be impaired.

A2. C204 is the neutralization feedback capacitor which cancels the effects of the collector-base capacitance. Without neutralization, the stage would tend to oscillate.

VIDEO AMPLIFIERS

Purpose: (1) Amplify the detected video information (a band of frequencies from 0 to 4 MHz) to a level sufficient to drive the crt. (2) Provide video output signals to various circuits in the receiver. (3) Provide a means whereby the viewer can control contrast and brightness. Some receivers also enable the operator to control the video frequency response. (4) In color receivers, delay the video signal to coincide with the chroma signal, causing each to arrive simultaneously at the crt.

Signals In: A composite video signal from the video detector. (See Fig. 4-10.)

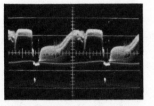

(A) *Vertical sweep rate.*

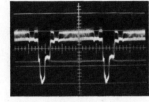

(B) *Horizontal sweep rate.*

Fig. 4-10. Composite video signal.

Signals Out: (1) Amplified composite video signal to drive the crt. (2) Amplified composite video signals to the sync separator, agc circuit, and chroma amplifier.

Circuit Description

Video amplifiers are wideband (0 – 4 MHz) class-A amplifiers which must produce sufficient voltage gain to drive the crt cathode (most sets produced at this time employ cathode drive). Because of the high peak-to-peak signal level required at this point, the video output transistor must be a high-voltage, high-power device. In early solid-state tv receivers, crts with low drive requirements were used. This somewhat reduced the requirements on the video output transistor. However, in the last several years, devices have been economically available which produce the approximate 150 volts peak-to-peak required to drive large-screen color crts.

Video amplifiers used in color receivers perform the same basic function as those in black-and-white receivers. However, there are some differences in the actual circuits.

The video amplifiers in color receivers invariably have more stages than those in black-and-white receivers. More stages are required because the three cathodes of a color crt represent a much greater load than the single cathode of a black-and-white crt. Some color receivers have as many as five video stages. Fig. 4-11 illustrates a simplified version of the video amplifier circuit used in a typical solid-state color receiver. Video amplifiers employed in color receivers are direct-coupled to assure proper matrixing of the video component with the color information at the crt (assuming matrixing occurs at the crt). Black-and-white video amplifiers may or may not be direct-coupled.

Color video amplifiers incorporate an additional component known as the delay line. The difference in bandwidths of the video and chroma information causes the transit times of the two signals through their respective circuitry to differ. The delay line causes the video information to be delayed so it will coincide with the chroma information. The delay line consists of a series of LC combinations, each of which provides delay of about 1 microsecond to the video information.

Most solid-state video amplifiers used in color receivers feature an automatic brightness-limiting circuit. The purpose of this circuit is to prevent sudden increases in crt beam current from overloading the horizontal-sweep circuits and possibly damaging the horizontal-output device or devices. Most of these circuits operate by sensing the pulse amplitude in the flyback circuit and converting this pulse to a dc level which controls the bias on one of the video amplifiers. If the beam current becomes too great, the direct-coupled video amplifiers are biased by the brightness-limiting circuit so as to reduce crt conduction.

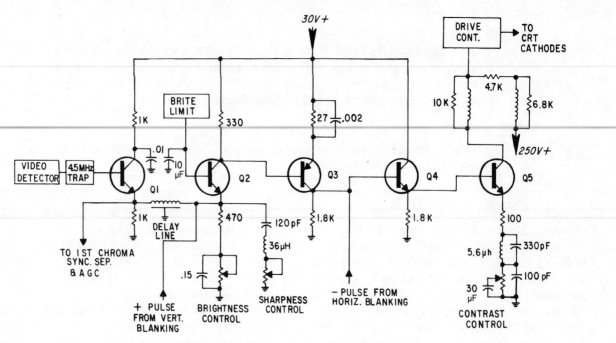

Fig. 4-11. Typical video amplifier used a solid-state color tv.

Most color receivers manufactured at this time matrix the chroma and the video information at the crt. The video amplifiers are about the same as shown in Fig. 4-11 with minor variations. Some manufacturers, however, matrix the chroma and video information just prior to the demodulator circuits. Sony and Motorola both matrix the video and chroma this way (Fig. 4-12). Each system has certain advantages over the other. The most obvious advantage of matrixing before demodulation is a reduction in the number of required video stages, since the color amplifiers located after the demodulators also function as video amplifiers.

The video amplifier illustrated in Fig. 4-13 is typical of solid-state, black-and-white tv receivers. Only two stages are required to supply the low-drive crt. The first video amplifier is an emitter follower, which is characteristic of both color and black-and-white video amplifiers. The emitter-follower configuration is selected to provide a sufficiently high input impedance to prevent loading of the video detector. Although an emitter follower is almost always used here, some receivers pick off signals at both the collector and emitter. The first video amplifier is usually the takeoff point for signals to other circuits such as agc, sync, and chroma. The low-impedance output of the emitter follower is direct-coupled to the video output stage.

High-frequency peaking is produced in the collector of the video output stage by the inductive and capacitive components present in the circuit.

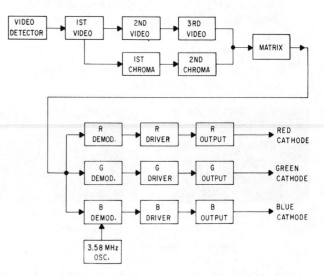

Fig. 4-12. Block diagram of chroma and video circuits which are matrixed before color demodulation.

Peaking is also achieved in the emitter circuit because of the value chosen for the emitter bypass capacitor. The capacitive reactance of 390 pF is such that significant bypassing occurs at the higher video frequencies. Therefore, the gain of the stage is greater at the higher frequencies.

The brightness is controlled by varying the dc level at the crt cathode. Maximum brightness is obtained when the control arm is at the ground end. The contrast control varies the amount of signal applied to the crt cathode. This particular receiver employs ac coupling through C268. The relatively high voltage required to supply the collector of the video output transistor and the crt

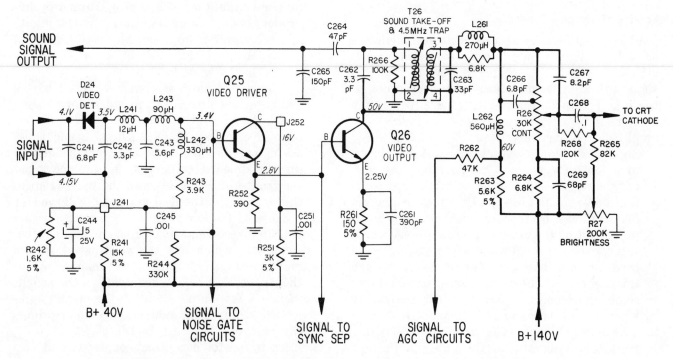

Fig. 4-13. Video amplifier used in typical solid-state black-and-white tv receiver.

cathode is often obtained by rectifing and filtering a portion of the horizontal retrace pulse from the flyback system.

Fig. 4-11 is a good representation of video amplifiers employed in most solid-state color receivers. Transistor Q1 is of the typical emitter-follower configuration, providing a high-impedance input to prevent loading of the video detector and a low-impedance output to drive other circuits in the receiver.

The output of Q1 is direct-coupled through the delay line to Q2, which is connected as a common-base amplifier. The common-base configuration is frequently used in video stages because of its inherently good high-frequency voltage gain. Since all the video stages are direct-coupled from the video detector to the crt cathode, brightness can be controlled by changing the dc operating point in any video stage. The conduction level of Q2 is varied by the brightness control without affecting the ac characteristics appreciably since the control is bypassed with a 0.15-μF capacitor. A positive pulse is applied to the emitter of Q2 which cuts it off during vertical retrace time, in turn cutting off Q3, Q4, Q5, and the crt. The frequency at the emitter of Q2 is selectively bypassed with a viewer-operated control. The values chosen will produce maximum bypassing and, therefore, maximum gain of Q2 at about 2.5 MHz. The base of Q2, being a common-base amplifier, is at signal ground. The dc bias at the base is determined by

the brightness-limiting circuit which senses the crt beam current and limits the conduction of Q2 accordingly.

Q2 is direct-coupled to the base of Q3, operating as a common-emitter amplifier. A pnp transistor is employed to allow direct coupling and to produce positive-sync signal polarity at the crt. (The sync signal at the video detector output is also positive.) Q4 is an emitter-follower stage providing a low-impedance source of current drive to the video output stage. A negative-going pulse is applied to the base of Q4, cutting it off during horizontal retrace time. Q4 in turn cuts off Q5 and the crt.

Video peaking is accomplished by selective bypassing of the high frequencies in the emitter circuit of Q5 and by emphasizing the high video frequencies in the collector circuit. The 10k and 6.8k swamping resistors lower the Q, and, therefore, provide the desired bandpass of the peaking inductors. The contrast control varies the stage gain, particularly at the lower video frequencies, by changing the degree of emitter degeneration.

Test Equipment

1. A vom or vtvm with a R×1 ohms range that does not produce a potential in excess of 1.5 volts at the test leads.
2. A jumper lead containing a 0.1-μF capacitor.
3. An oscilloscope. Low-capacity probes are useful in some stages.

Service Hints and Procedures

Video amplifiers, being low-frequency amplifiers as well as high-frequency amplifiers, lend themselves very well to simple signal-tracing and signal-injection techniques. A capacitor probe of the type shown in Fig. 1-1 is generally the only equipment required to isolate a defect in the video amplifier to a particular stage.

1. Signal injection. Connect the capacitor probe to a low-voltage 60-Hz ac source. Most low-voltage supplies (12 to 30 volts) before rectification are suitable signal sources or the crt filament may be more convenient. Touch the probe end of the capacitor probe to each base successively. Normally operating stages should produce a hum bar across the crt face. Although this method is effective in most cases, it has some shortcomings and may be misleading when certain types of defects are involved.

2. A somewhat more effective troubleshooting method is to signal trace the air signal through the video stages. Select a known operating channel and tune for best sound if present. Connect the clip end of the capacitor probe to the base of the first audio amplifier. Alternately probe the output of each stage. A 60-Hz sync buzz should be heard if the stage is operating. The relative stage gain, as indicated by volume changes, is somewhat more easily determined by this method than by the 60-Hz signal-injection technique.

3. Each of the preceding techniques is quite useful, particularly to the outside service technician. However, a certain degree of experience is required, since some objective evaluation of the test results is required to be meaningful. The best troubleshooting method is to signal trace the air signal with an oscilloscope. The quality as well as the amplitude of signal can now be observed. If a triggered-sweep scope is available, the frequency response of the video amplifiers can even be determined by viewing the VIT pulse (vertical interval test signals).

4. Service technicians accustomed to servicing tube-type circuits lean toward troubleshooting strictly by voltage measurements. In direct-coupled transistor circuitry, this is about the most confusing method that can be employed. For example: A defect in the video detector stage (Fig. 4-11) can cause all the voltage readings in the video output stage (Q5) to be off. By the time you logically determine that improper voltages on Q5 are being caused by improper conduction of Q4, caused by Q3, due to Q2, as a result of improper bias on Q1, because of a defect in the video detector circuit, a mental hernia may develop.

Inoperative video is usually a result of the failure of one or more transistors. (More than one is the rule rather than exception in direct-coupled circuits.) A most effective and time-saving troubleshooting practice, particularly in direct-coupled circuits, is to simply check out the front-to-back ratio of all the transistors in the circuit. Use the R×1 scale of the ohmmeter because of the low shunting resistances occurring in direct-coupled circuits. Open or shorted junctions, which are the most usual failure modes, are easily located in this manner. All of the devices can be checked while in the circuit in just a few minutes and all the mental gymnastics required for analyzing voltage readings can be avoided.

5. Each transistor in a direct-coupled circuit acts as the control for each successive stage. This allows some simple checks concerning the dc operation of the various stages. For example, cut off Q1 in Fig. 4-11 by shorting the base to the emitter. The conduction of Q2 should increase and Q3, Q4, and Q5 should conduct more, causing the crt brightness to increase. If this does not occur, a stage between Q1 and the crt is inoperative. Short the emitter to the base of Q2. Q2 should cut off and Q3, Q4, and Q5 should also cut off, causing minimum crt brightness. If minimum brightness does not occur, a stage between Q2 and the crt is inoperative. Each stage can be disabled successively in this manner until the offending stage is located.

6. Delay-line defects usually cause no video (and no raster in most sets) or ringing (multiple images). The multiple images do not change with antenna orientation and will change only slightly with fine tuning. Three delay-line malfunctions are possible. They are: an open coil, coil shorted to ground, or an open ground.

A coil which is open or shorted to ground will result in no video and, in most receivers, no raster. An open ground connection will cause ghosting or ringing. If a defective delay line is suspected, simply disconnect both signal ends of the delay line and bridge the open circuit with a jumper lead. Restoration of the picture or elimination of the ringing indicates a defective delay line.

REVIEW

Q1. What is the purpose of C251 in Fig. 4-13?

Q2. Is the composite video signal audible?

Q3. Should a detector probe be used when signal tracing in the video amplifier circuits?

Q4. What is the purpose of C261 in Fig. 4-13?

Q5. Why is a special brightness-limiting circuit used in all solid-state color receivers?

Q6. Is retrace blanking always accomplished in the video amplifier?

Q7. Is the cathode or anode of the video detector diode connected to the 4.5 MHz trap in Fig. 4-11? Does it matter?

Q8. What symptoms would be apparent if Q1 in Fig. 4-11 shorts from emitter to collector?

Q9. What would occur if the 10-μF capacitor in the base circuit of Q2 in Fig. 4-11 opened?

Answers

A1. It is customary to place the collector at signal ground in emitter-follower circuits. Emitter-follower circuits tend to oscillate if this is not done.

A2. Yes. The 60-Hz vertical sync pulse is audible as a 60-Hz buzz and can be used for signal tracing.

A3. No. The signal has already been detected by the video detector.

A4. C261 provides selective bypassing of the higher frequencies to increase gain at the higher video frequencies (peaking).

A5. The function of the brightness-limiting circuit is primarily to protect the horizontal output device or devices from sudden overloads due to crt beam current loading of the flyback system.

A6. No. Some receivers accomplish blanking in the color-difference amplifiers. Receivers which matrix the video and the chroma before demodulation cut off the crt during retrace by controlling conduction of the demodulators.

A7. The cathode. The video information must be of sync-positive polarity from the detector to produce sync-positive video at the cathode of the crt. If the diode were reversed, a negative picture could result, in addition to the sync and agc operating improperly. The polarity of the video detector depends on the configuration and number of video stages. The polarity of the video signal at the crt cathode must always be sync-positive.

A8. Brightness will diminish, probably to the point of no raster. Of course, the video information will also be absent.

A9. The gain of Q2 would be drastically reduced. Q2 is operating as a common-base amplifier. The 10-μF capacitor is in the input signal path of the amplifier.

CHAPTER 5

Chroma Amplifiers

Chroma or color amplifiers can be roughly divided into two general types; those which amplify the signal before demodulation and the amplifiers employed after demodulation. The chroma amplifiers employed before the demodulators may be considered as i-f amplifiers since the bandwidth of the chroma circuits is determined by these stages. The tuned chroma circuits produce a bandwidth of about 1 MHz at a center frequency of 3.58 MHz.

The color amplifiers employed after demodulation may be considered as video amplifiers. In most receivers these stages must amplify frequencies from about 0 to 1 MHz. In receivers which matrix the video and the chroma signals before demodulation, these stages must be capable of amplification to 4 MHz, since they truly become video as well as color amplifiers. The amplifiers which occur in the circuit after demodulation are usually termed color-difference amplifiers if video and color information are matrixed at the crt. If matrixing occurs before demodulation, the stages are simply called color amplifiers; R, G, and B amplifiers; or color video amplifiers.

Solid-state receivers usually employ two or three stages of chroma amplification before demodulation and one or two stages of color amplification after demodulation. Control functions such as tint or hue, color intensity, and color-killer action take place before demodulation.

CHROMA AMPLIFIERS
BEFORE DEMODULATION

Purpose: (1) Amplify the chroma sideband information consisting of frequencies from 3.08 MHz to 4.08 MHz. (2) Separate chroma signals from the composite video information. (3) Provide variable gain characteristics to maintain relatively constant chroma level (acc). (4) Provide a means of adjusting tint and chroma gain. (5) Provide a means of disabling the chroma circuits during black-and-white transmission. (6) Provide a means of keying burst from the chroma signal and at the same time preventing the burst signals from reaching the demodulators.

Signals In: (1) Composite video (Fig. 5-1A). (2) A dc level which increases in the positive direction, as burst and chroma increase in amplitude, to control the gain of one of the chroma amplifiers (acc). (3) A dc level which cuts off one of the chroma amplifiers during black-and-white transmissions. (4) A retrace pulse, in some receivers, to cut off the final chroma amplifier during burst time (Fig. 5-1B). A positive or negative pulse may be used depending on the type of device and circuit configuration used.

Signals Out: (1) Amplified chroma information with the burst signal keyed out (Fig. 5-1C). Most receivers key the burst signal out from the final chroma amplifier. The Silvertone chroma circuit shown in Fig. 5-2 does not remove burst from the composite chroma prior to demodulation. Rather, the demodulators are cut off during retrace time by removal of the 3.58-MHz oscillator signal. Since burst occurs during retrace, it is keyed out at the demodulators. (2) Composite chroma (Fig. 5-1D) to the burst keyer or amplifier stage.

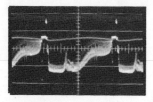

(A) Sync-positive composite video.

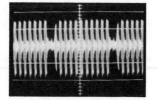

(B) Negative-going horizontal retrace pulse.

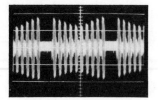

(C) Chroma with burst signal keyed out.

(D) Chroma with burst.

Fig. 5-1. Waveforms present in the chroma circuits.

Circuit Description

Fig. 5-3 illustrates a simplified version of the chroma amplifier used in RCA solid-state color receivers. Essentially, three stages of chroma amplification are provided, with Q3 functioning as a phase-shift circuit for control of tint. Generally, more stages are required in solid-state receivers. The functioning of the two circuits shown in Figs. 5-2 and 5-3 are much the same with minor variations.

The input or chroma takeoff circuit is a series-resonant circuit which allows the high-frequency component of the video signal (chroma information) to reach the base of the first chroma amplifier. The chroma takeoff coil is tuned to the high-frequency end of the response curve, around 4.08 MHz, to compensate for the reduced gain of the higher-frequency chroma signal in the i-f amplifiers.

The chroma amplifiers are class-A, capacity coupled, high-gain voltage amplifiers. The amplifier stages in Fig. 5-2 are both fully bypassed for maximum gain. Q2 and Q4 in Fig. 5-3 have some emitter degeneration due to the unbypassed 56-ohm and 47-ohm emitter resistors.

The bias on the base of the first chroma amplifier in both Fig. 5-2 and Fig. 5-3 is derived from a separate circuit; the acc (automatic color control) amplifier. This is a dc amplifier which develops a voltage that becomes increasingly positive as the amplitude of burst signal increases. The positive-going dc voltage applied to the npn transistor causes increased conduction, which decreases gain (forward agc action). Transistors operated in a forward-agc mode are specially designed to produce a fairly linear reduction in the high-frequency beta characteristics with increased emitter current. In addition to the gain reduction achieved in this manner, the transistor in Fig. 5-2 loads the tuned circuit somewhat, causing lower Q and a further reduction in gain. The bandpass of the chroma circuits is determined by the double-tuned response of the chroma bandpass transformer. The tuned circuits are placed between the first and second chroma amplifiers in Fig. 5-2 and in the output stage in Fig. 5-3.

Tint or hue can be controlled by shifting the phase of the chroma information or the phase of the 3.58-MHz oscillator signal fed to the demodulators. The examples in Figs. 5-2 and 5-3 illustrate how to achieve tint control by shifting the chroma phase. The circuit shown in Fig. 5-2 selects either of two degrees of phase shift depending on the position of a viewer-operated switch (ATL switch). The purpose of this circuit is to reduce the amount of noticeable flesh tone changes with varying signal and when changing channels. With the ATL switch in the off position, R86 is shorted to ground and the tint control R88 operates in the normal manner. With the ATL switch in the on position, the chroma signal is phase shifted a fixed amount by R91 and L82. At the same time, R86 is connected into the tint control circuit, reducing its effect. A second section of the ATL switch (not shown) shorts out an inductance in the X demodulator, increasing the phase angle between the X axis and the Z axis. This increase in phase angle broadens the range of flesh tones.

The circuit shown in Fig. 5-3 incorporates a rather unique tint-control circuit. A transistor (Q3) is employed to produce the required phase shift. In the common-emitter configuration, the emitter and collector signals are 180° out of phase. The chroma signal is taken off at both the emitter and collector. The degree of phase shift is determined by the amount of signal from the collector which is established by the setting of the tint control. One advantage of such a system is that the phase can be easily shifted over a broad range with virtually no change in amplitude.

The bias on Q4 is obtained by conduction through the color killer transistor. When no burst is present, the killer transistor does not conduct and the chroma output stage is disabled. A negative horizontal retrace pulse is also applied to the base of Q4 to cut it off during the period of time that burst occurs. This keys the burst signal out of the chroma information applied to the demodulators.

The color killer function of the amplifier in Fig. 5-2 takes place in the second chroma stage and

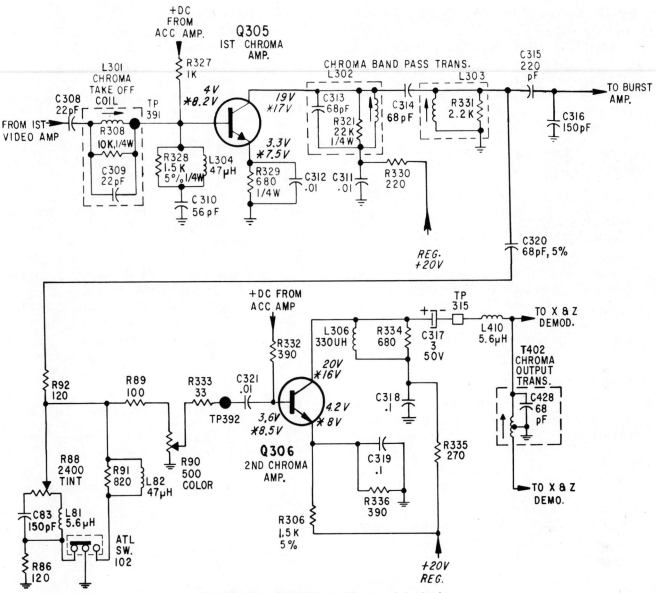

Fig. 5-2. Chroma amplifier in Silvertone hybrid color tv.

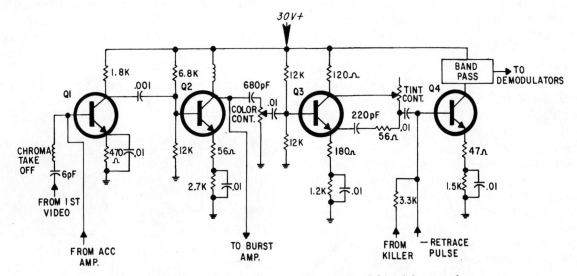

Fig. 5-3. Simplified version of chroma amplifier used in RCA solid-state color tv.

is produced by the voltage from the acc amplifier. A positive voltage is applied to the emitter of Q306 from the divider R306 and R336. The voltage from the acc amplifier which is applied to the base of Q306 is not sufficiently positive to forward bias the transistor when receiving a black-and-white program. However, when burst is present, the acc voltage rises several volts and causes the stage to conduct.

Test Equipment

1. A capacitor jumper lead containing a 0.01-μF capacitor as shown in Fig. 1-1.
2. A vom or vtvm.
3. An oscilloscope with a detector probe and a low-capacity probe.
4. A sweep and marker generator with marker frequencies at 4.08, 3.58, and 3.08 MHz.
5. A color-bar generator. This is useful for providing a color signal in the absence of color transmission. However, at the present time, such situations would be quite rare.

Service Hints and Procedures

1. All color receivers employ a color-killer circuit to disable the chroma amplifiers during black-and-white transmission. Most present-day receivers employ an automatic color gain control circuit (color agc). In cases of no or weak color, four steps should be taken as follows:

   (a) Check the fine tuning.
   (b) Check setting of the color intensity control.
   (c) Disable the color killer circuit.
   (d) Disable the color agc or acc circuit.

   Color killer and acc circuits vary from one receiver to another and each circuit requires a specific means for defeating it, usually described by the manufacturer. Initially, the setting of the color-killer control should be checked by rotating it from one extreme to the other. Although this is not conclusive check or a positive means of disabling the color killer, it will determine if the control is simply misadjusted and killing the color. In Fig. 5-3, the color killer can be defeated by shorting the collector of the color-killer switch transistor to the emitter. This allows normal conduction of Q4. In this same circuit, the acc is defeated by shorting the base of the acc amplifier to the emitter. The transistor cuts off, causing the collector voltage to rise to maximum. Q1 then operates at maximum gain. Some chroma circuits, such as illustrated in Fig. 5-2, required that a specific bias voltage be applied as a substitute for the

acc voltage which, in this case, acts as gain control and provides color-killer action.

2. No or weak color may also be caused by misalignment. Alignment is quite simple and most receivers require only three adjustments: the takeoff coil and the two bandpass transformer slugs. A detector probe must be used with the oscilloscope. Follow the specific instructions given by the manufacturer for alignment.

3. The most effective troubleshooting method is to signal trace a color signal, from the air or a color-bar generator, through the amplifiers with an oscilloscope. A low-capacity probe or some such isolation probe is required to prevent loading. Most chroma stages are high-gain voltage amplifiers and appreciable gain should be observed in each stage. Most receivers will exhibit a substantial reduction of signal amplitude through the bandpass transformer. This is normal since the transformer is usually coupling a high-impedance output to a low-impedance input. (Voltage gain is traded for current gain.)

COLOR AMPLIFIERS AFTER DEMODULATION

Purpose: (1) Amplify the color signals consisting of frequencies from approximately zero to 1 MHz to a level sufficient to drive the crt. (2) Develop the green signal (some receivers).

Signals In: (1) Chroma signal (Fig. 5-4A). (2) Horizontal retrace pulse for retrace blanking and dc restoration (Fig. 5-4B).

Signal Out: Amplified color information.

Circuit Description

The color amplifiers employed after demodulation are of three different basic types. Fig. 4-12 is a block diagram of the system used by Sony and Motorola. The color amplifiers process both the color and the video signals simultaneously. The other two systems are quite similar to each other, differing only in the method by which the green signal is developed. In either case, the amplifiers

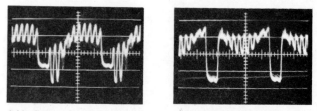

(A) *Color signal at one of the crt grids.*   (B) *Negative horizontal retrace pulse.*

Fig. 5-4. Waveforms in the chroma output circuits.

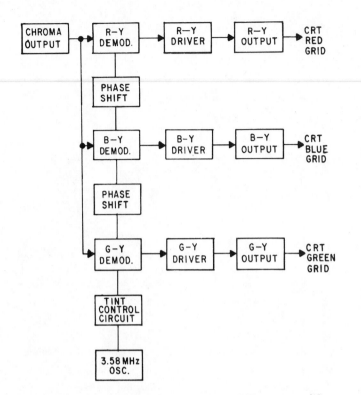

Fig. 5-5. Block diagram of a typical color difference amplifier.

are referred to as color difference amplifiers since only the color signals are amplified. The Y or video signal is matrixed with the color at the crt.

Fig. 5-5 is a block diagram of the color difference amplifiers used in typical receivers by RCA and others. All three colors are demodulated by separate demodulator circuits. The 3.58-MHz oscillator reference signal is properly phase shifted at each demodulator to produce the correct color.

Fig. 5-6 illustrates color difference amplifiers which develop the G − Y signal by combining the correct amount of R − Y and B − Y signals. The G − Y amplifier is operated as a modified common-base amplifier. The 3.9k resistor in the base circuit raises the input impedance of the G − Y amplifier to more closely match and balance the operation of all three stages. A negative horizontal retrace pulse is applied through diodes to each crt grid. This pulse provides dc restoration due to

the charge-discharge time of the 0.01-μF coupling capacitors. The proper conduction level of the three guns is thereby restored, regardless of the instantaneous frequency of the color signal. When dc restoration is accomplished in this manner, the need for dc coupling between the collectors of the color-difference amplifiers and the crt grids is eliminated. In some receivers, horizontal retrace blanking also takes place at this point. However, in the circuit shown in Fig. 5-6, blanking occurs by cutting off the video amplifier.

Test Equipment

1. A capacitor jumper lead containing a 0.1-μF capacitor as shown in Fig. 1-1.
2. A vom or vtvm.
3. An oscilloscope.

Service Hints and Procedures

1. Since the color stages amplify the low frequencies as well as the high frequencies, signal injection becomes very simple. A 60-Hz signal injected with the capacitor probe into the suspected stage or stages will produce a hum bar with the particular color characteristics (red, blue, or green) of the circuit.
2. Problems which affect both color and crt conduction (low brightness, excessive brightness, blooming, or in some receivers, vertical retrace lines) are usually associated with the color difference amplifiers. In most receivers, the retrace pulses applied to these stages establish the average conduction of each gun. Any component failure which disturbs the forming or generation of this pulse also affects crt conduction and the color.
3. Signal tracing a signal from the air or a color-bar generator with an oscilloscope is good shop procedure. A direct probe can be used in these circuits.
4. When crt conduction problems are apparent, the gray-scale setting should be checked. Each manufacturer supplies specific instructions regarding this procedure.

REVIEW

Questions

Q1. What are the two functions of the chroma takeoff circuit?

Q2. The center frequency of the chroma signal is 3.58 MHz. Therefore, a detector probe is not required when signal tracing the air signal with a scope. True or false?

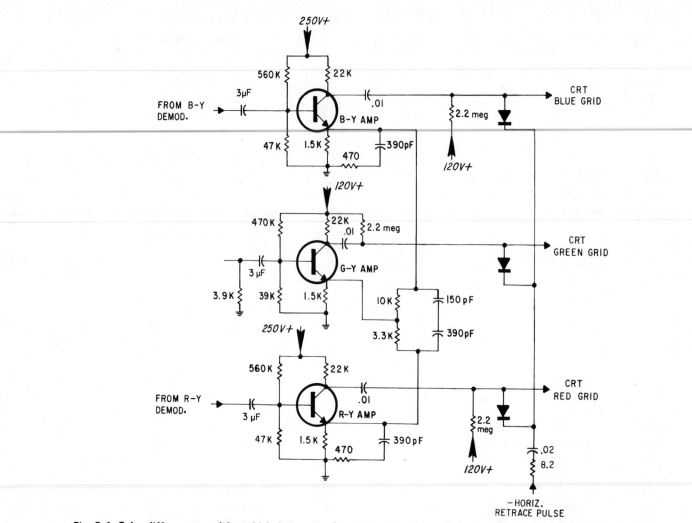

Fig. 5-6. Color difference amplifier which derives the G — Y signal in the emitter circuits of the R — Y and B — Y stages.

Q3. A detector probe is required when aligning the chroma circuit. True or false?

Q4. What are the two methods of controlling tint?

Q5. Why are the emitter resistors of the R – Y and B – Y amplifiers in Fig. 5-6 not fully bypassed?

Q6. If both color and brightness are improper, which of the components in Fig. 5-6 would most likely be at fault?

Answers

A1. The circuit is tuned to pass the higher video frequencies which include the chroma information. The lower-frequency video information which includes the sync pulses is excluded. The circuit is aligned toward the high end of the chroma frequencies (4.08 MHz) to compensate for the i-f response which accentuates the low end (3.08 MHz).

A2. True. Modern scopes intended for color work have sufficient response to view the 3.58-MHz chroma signal directly.

A3. True. A response curve can only be developed from a detected signal. The detector quadrupler illustrated in Fig. 1-2 is also suitable.

A4. The phase of either the 3.58-MHz reference signal or the chroma signal can be shifted before it is applied to the demodulators.

A5. The signal applied to the G − Y amplifier is developed across the emitter resistors of the R − Y and B − Y amplifiers. Complete by-passing would, as a result, be impossible in such a circuit configuration. Some receivers pick the signals off the collector of the B − Y and R − Y amplifiers and feed the base of a common-emitter G − Y amplifier. While each system has certain advantages, emitter feed is usually simpler. This is because the low impedances involved are compatible with good high-frequency response without additional tuning or peaking.

A6. The diodes. A defective diode will cause loss of dc restoration and improper crt grid voltages.

CHAPTER 6

Audio Amplifiers

Solid-state audio amplifiers are of many different circuit configurations and degrees of complexity depending on application and price range. Furthermore, the dual nature of bipolar transistors (npn and pnp) allows circuit designs which are impossible with vacuum-tube devices. In this chapter, we will discuss the four most popular amplifier types now in use. They are: Class-A transformer-coupled, class-B transformer-coupled, class-B direct-coupled complementary-symmetry, and class-B direct-coupled quasi-complementary-symmetry. The last three types are actually class-AB amplifiers as will be explained subsequently in this chapter.

CLASS-A TRANSFORMER-COUPLED AMPLIFIERS

Figs. 6-1 and 6-2 are typical of class-A amplifiers used in ac-operated equipment. Both examples are used in portable phonos. However, a version of the amplifier shown in Fig. 6-1 is also quite common in tv receivers. Battery-operated equipment seldom employs class-A audio amplifiers because of the high no-signal current drain. An exception to this is an auto radio where current drain is not a particularly critical factor.

Notice in Fig. 6-1 that the bias for the driver stage is derived entirely from the conduction of the output stage, producing a voltage drop across R212. The two transistors are operating in a closed-loop configuration which stabilizes the operating point of the output transistor. As the temperature rises, the conduction of the output transistor tends to rise and the drop across R212 increases. This increases the forward bias and the conduction of the driver transistor. The collector voltage of the driver stage then falls, reducing the conduction of the output stage. Because of dc

coupling and feedback, the entire circuit acts as a single device and must be viewed as such. A defect in either stage causes both stages to operate improperly.

R215 is a voltage-dependent resistor which protects the output transistor from voltage spikes reflected from the output transformer. Any transient audio noise of sufficient amplitude and rise time to cut off the output stage will produce a high-voltage kickback pulse due to the rapidly collapsing field in the primary of the transformer. The voltage-dependent resistor conducts and shunts the pulse, preventing it from appearing at the collector of the transistor.

Fig. 6-2 is a higher-power class-A phono amplifier. The output stages and driver stages are stacked or in series, which is common in class-AB push-pull amplifiers but somewhat unusual for class-A types. The operating point of the output devices is also stabilized by a closed-loop, dc-feedback system. The midpoint voltage at A will be equal to one half the supply voltage if both output transistors are biased at the same operating point. This midpoint voltage is fed back to set up the operating point of the driver transistors. Therefore, the two driver transistors and the two output transistors in this circuit must each be considered as a part of one single closed-loop system. A malfunction of any one of the transistors will cause improper voltages to appear in the three remaining stages.

Most class-A amplifiers are transformer coupled to the speakers. In this case, matching is provided by autotransformers. Capacitor C102B provides dc isolation for the left speaker. No isolation is required for the right-channel speaker, since the small dc flow through the speaker at this point is negligible.

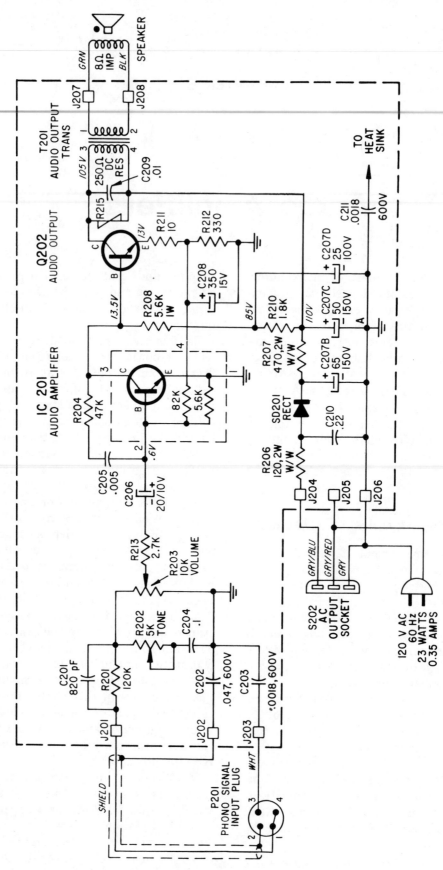

Fig. 6-1. Typical class-A phono amplifier.

110

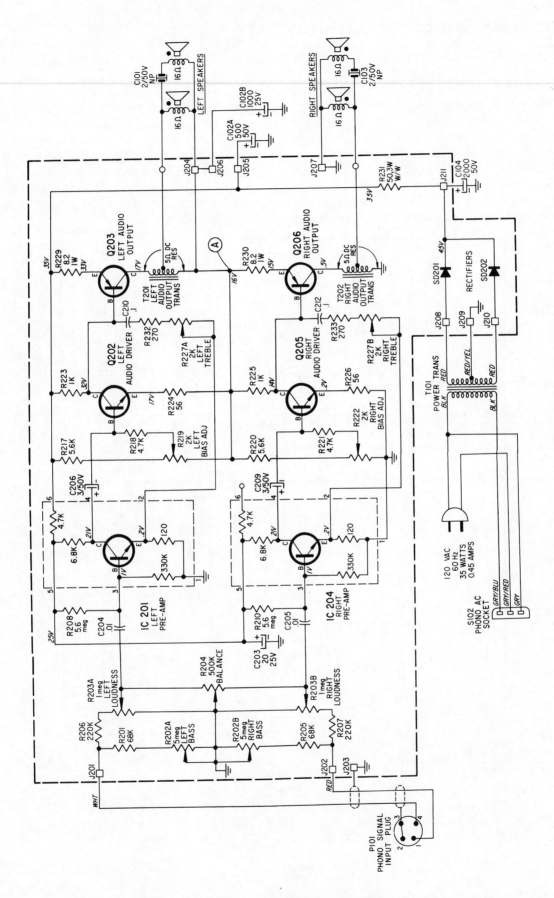

Fig. 6-2. Higher-power class-A phono amplifier.

CLASS-B TRANSFORMER-COUPLED AMPLIFIERS

The amplifier illustrated in Fig. 6-3 is representative of the output stages of high-power (10 watts and higher), push-pull, class-AB amplifiers used in console and component stereo units. Very low power amplifiers of this configuration are also popular in portable radios. The circuit design in such cases is similar, although the component values differ widely. Amplifiers of this type are usually referred to as class-B amplifiers. In actual practice, however, transistor audio amplifiers are never operated class-B because crossover distortion will result. This distortion arises as a consequence of the inherent emitter-base diode voltage drop. The transistor is not forward biased until this voltage is overcome. Forward bias is therefore applied, causing a constant, small, quiescent (no-signal) current to flow. The operation, strictly speaking is class AB. Transformer-coupled amplifiers of this nature are easier to service since the operating points of the output transistors are independent of the other stages. Output transistor failure is also the most likely malfunction in this type of amplifier due to the high power levels being handled.

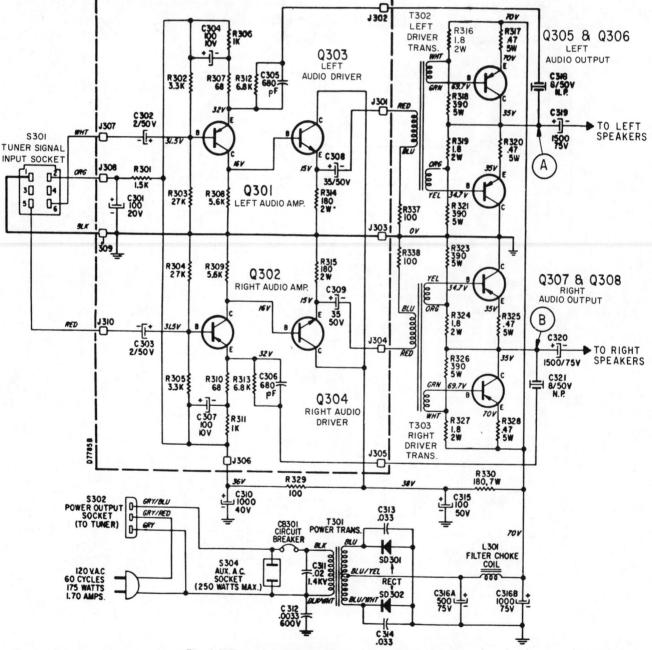

Fig. 6-3. Typical class-B, transformer-coupled amplifier.

112

Stabilization of the output transistors is a result of the values chosen for the emitter and biasing resistors. A stability factor of about 4 is obtained with the .47-ohm emitter resistor and the 1.8-ohm base resistor. Ideally, the midpoint voltage at points A and B should be one half the supply voltage. Most defects in the output stage will cause this voltage to deviate from its normal value. Therefore, measurement of the midpoint voltage is a good check to determine if the output stage is operating correctly.

The amplifier illustrated in Fig. 6-3 employs a negative-ground power supply. Many amplifiers of this type incorporate a split-polarity power supply that produces equal negative and positive potentials in relation to chassis ground. When such an arrangement is used, the dc midpoint voltage is zero relative to ground and the large speaker-coupling capacitors (C319 and C320) can be eliminated. A degenerative ac feedback loop from the output stages to Q301 and Q302 stabilizes the operation of the amplifier. The driver stages are direct coupled and must be viewed as a single device regarding dc operation. Notice that Q303 and Q304 are of emitter-follower configuration and, therefore, no voltage gain will be apparent in these stages.

CLASS-B DIRECT-COUPLED AMPLIFIERS

The dual nature of transistors allows simple push-pull amplifier circuits without the necessity of power transformers for phase inversion. These circuits are called *complementary-symmetry* amplifiers. Two such amplifiers are depicted in Figs. 6-4 and 6-5. The theory of operation is the same in each circuit, although the power-output capabilities differ. Fig. 6-4 is a low power (1 watt) portable phono amplifier, while the circuit Fig. 6-5 is used in a console unit and has a power output of approximately 7 watts. These amplifiers are also, strictly speaking, class-AB amplifiers, as a no-signal current is required to prevent crossover distortion. Only the left channel of each amplifier is illustrated. An identical right channel is, of course, utilized for stereo operation.

The three output transistors in Fig. 6-4 and the four transistors in Fig. 6-5 are direct coupled. The transistors act as a single device and any malfunction in one transistor reflects a change in the dc operating point of the others. As a result, techniques other than voltage measurements are usually more effective when servicing such circuits.

The operation of complementary-symmetry amplifiers is most easily understood when considered in terms of dc paths. Therefore, the biasing circuits of the output stages should be examined. The biasing circuits are quite simple if redrawn in terms of an equivalent circuit where the driver transistor is shown as a variable resistance. Fig. 6-6A illustrates the equivalent biasing circuit for the amplifier in Fig. 6-5. Fig. 6-6B is the equivalent circuit for Fig. 6-4.

The biasing circuits, when redrawn, can be seen as essentially simple divider circuits. The current through each of the bias divider networks is controlled by its respective driver transistor (Q110). The quiescent current, as well as the instantaneous current which changes with signal, determine the bias to the output transistors. As Q110 in Fig. 6-6A increases conduction, the bias voltage fed to each output transistor will become

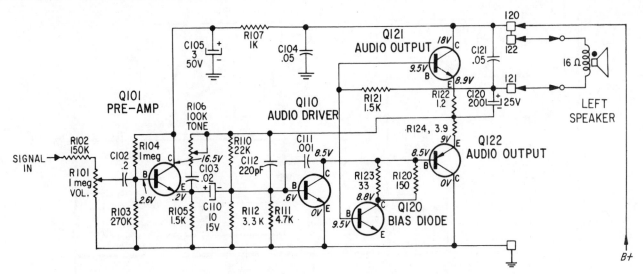

Fig. 6-4. Low-power, class-B, direct-coupled amplifier.

more positive. The difference between the two bias voltages will remain essentially constant due to the regulating action of D110. As a more-positive voltage is applied to Q120, its conduction will *increase*. This same positive bias increase is also applied to Q121, which causes its conduction to *decrease* a like amount. As driver current decreases, the opposite action occurs. The midpoint voltage between the output transistors will, therefore, rise and fall according to the applied signal. The fluctuating voltage at this point causes the speaker coupling capacitor C120 to charge and discharge, creating an alternating or push-pull current through the voice coil of the speaker. The same basic action occurs in the circuit shown in Fig. 6-6B. However, in this case, increased conduction of the driver transistor causes the bias

voltages to diminish or go negative. These changing bias voltages also cause the npn-pnp output pair to react oppositely.

The dc operating point of the direct-coupled portion of the circuits in Figs. 6-4 and 6-5 is determined by a closed-loop feedback system in conjunction with the stabilizing effect of emitter feedback and the biasing diodes. Diodes (a transistor junction in Fig. 6-4) are used in the biasing circuits of the output transistors because of their temperature characteristics. As the temperature rises, the voltage drop across the diode or diodes decreases, reducing the bias to the output devices.

When the output transistors are biased correctly at the same operating point, the midpoint voltage will be one half the supply voltage. In Fig. 6-5, the midpoint voltage is fed back to the

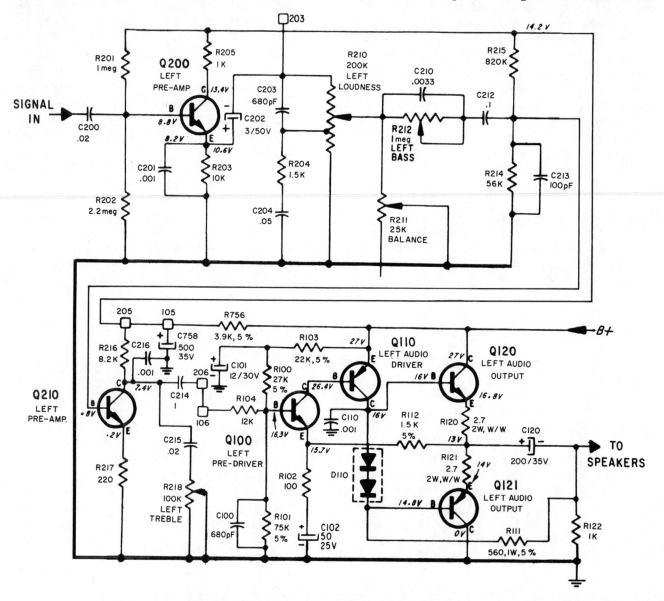

Fig. 6-5. Higher-power, class-B, direct-coupled amplifier.

emitter of Q100, the predriver. The conduction of this transistor determines the conduction of the driver Q110, which controls the operating point of the output devices. Consider, for example, the effect of a rising midpoint voltage. A more-positive voltage on the emitter of Q100 causes less conduction which, in turn, reduces the conduction of Q110. The collector voltage of Q110 will, as a result, decrease. A decreasing voltage at this point will cause Q121 to be more forward biased. Originally, a rising midpoint voltage could have only been a result of the opposite biasing condition. A more-negative or falling midpoint voltage will produce the reverse effect.

The amplifier in Fig. 6-4 also makes use of the midpoint voltage to create a closed-loop stabilizing system. In this case, the midpoint voltage determines the base bias of Q110. A rising midpoint voltage causes increased conduction of Q110, which forward biases Q122 more and Q121 less. The normal operating level of the output transistors is thereby restored.

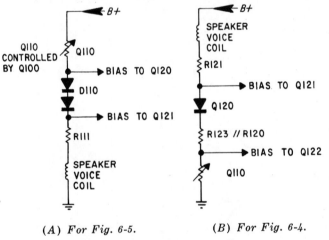

(A) For Fig. 6-5.   (B) For Fig. 6-4.

Fig. 6-6. Equivalent biasing circuits.

The ac operation of the amplifier is also stabilized by means of this feedback loop, since the signal as well as the dc level is fed back in a degenerative manner. The speaker voice coil, being a part of the biasing network, also provides some ac feedback.

The preamplifier stages of both amplifiers are capacitively coupled and, as such, do not pose any unusual service problems. Normal signal tracing or injection techniques will usually isolate problem stages.

High-power audio amplifiers (10 watts and greater) are usually either transformer coupled or of the so called *quasi-complementary-symmetry* configuration as illustrated in Fig. 6-7. The term *quasi* merely indicates that the complementary-

symmetry portion of the circuit is before the output stage. Matched complementary pairs of high-power transistors are relatively expensive. Incorporating the complementary pair before the output stage provides the phase inversion required to drive the pnp output transistors and reduces the cost at no sacrifice to performance. The five transistors after the loudness control in each channel are all direct coupled. Failure of any one of the transistors will cause the operating point of the remaining four to shift. All voltage readings throughout the entire circuit will, as a result, differ from the normal operating voltages indicated on the schematic.

Stability is achieved by closed-loop feedback in a similar manner to the circuits previously described. The midpoint voltage between the stacked output transistors will again be one half the supply voltages when operating properly. The midpoint voltage is fed back to set up the emitter voltages of the complementary pair and the base voltage of the driver. Considering the right channel, a rising midpoint voltage will cause Q210 to be biased less, Q209 more, and Q206 more. The collector voltage of Q210 will, therefore, rise and the emitter voltage of Q209 will fall. Q214 will, as a result, be biased on less and Q213 on more, bringing the midpoint voltage back to one half the supply voltage and stabilizing the operation of the circuit.

The equivalent biasing circuit for the complementary-symmetry transistor pair is as shown in Fig. 6-8. Notice that the input circuits of the complementary pair are not symmetrical. The 1.2k resistor R240 balances out this mismatch. The feedback loop in this amplifier is primarily of a dc nature. The ac component is filtered by C226 and R236.

Test Equipment

1. A capacitor jumper lead containing a 0.1-μF capacitor as illustrated in Fig. 1-1.
2. A vtvm or vom.
3. An oscilloscope.
4. An audio sine-wave and square-wave generator. Most problems can be solved just as quickly without the use of either a scope or a generator. At times, however, certain problems such as distortion may require the use of a generator. Most present-day amplifiers are dual channel (stereo) which greatly eases servicing through signal-jumping and comparison techniques.
5. A battery-operated audio amplifier is very useful for signal tracing. Several low-cost versions are on the market.

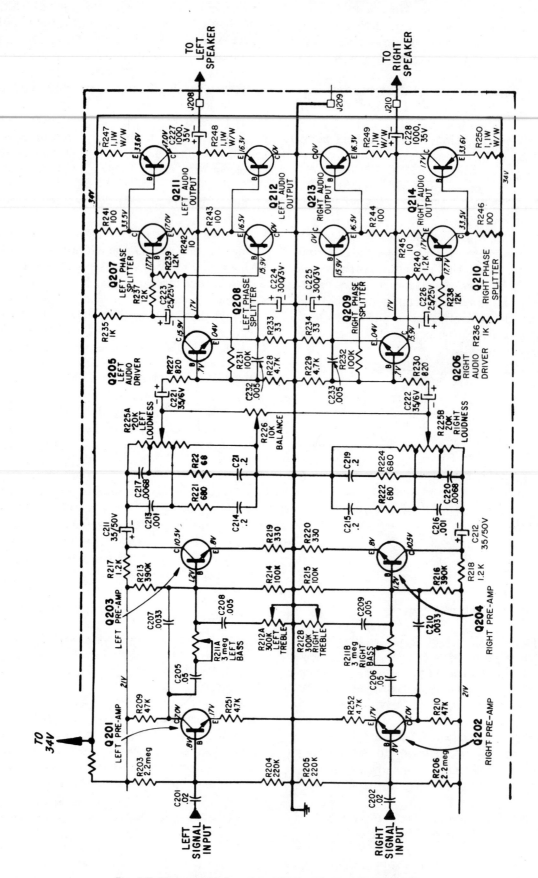

Fig. 6-7. Direct-coupled, quasi-complementary-symmetry amplifier.

116

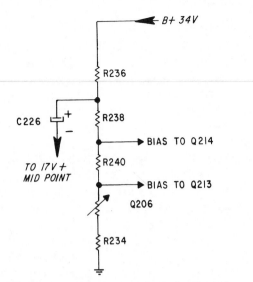

Fig. 6-8. Equivalent biasing circuit for Fig. 6-7.

Service Hints and Procedures

1. The operating point of each transistor in a direct-coupled circuit hinges on the operation of all the others. Isolation of a defective stage cannot, for this reason, be easily accomplished by voltage measurements as in capacitively coupled stages. The most common defect in such amplifiers is transistor failure in terms of open or shorted elements. The simplest and most effective troubleshooting method is an in-circuit check of *all* the transistors in the direct-coupled portion of the circuit. Notice that *all* is emphasized. Many technicians experience difficulty with direct-coupled circuits because, when locating a defective device, they will assume the problem has been solved and proceed to replace the defective transistor. Multiple transistor failure is the rule rather than the exception in direct-coupled amplifiers. All the transistors should be checked before applying power.

   With the power disconnected, check the front-to-back ratio of both junctions and the emitter-to-collector resistance of each transistor. This can be accomplished in circuit by using the R×1-ohm scale of the meter. (The meter must not produce lead potentials in excess of 1.5 volts.) The R×1 scale should be used because of the low parallel resistances found in a direct-coupled circuit. Open or shorted transistor elements are easily isolated by this technique and the majority of dead units will be repaired more quickly by this method than any other.

2. Strict attention must be paid to transistor basing, particularly if replacements are from a different manufacturer than the original. This cannot be stressed strongly enough. Improper installation of transistors is very easy and an extremely common error on the part of even experienced technicians. Reversed emitter and collector leads will, in many cases, result in an operating amplifier. However, it will not function properly.

3. In direct-coupled, stacked-output amplifiers, the most significant voltage measurements is the midpoint voltage between the two output transistors. When the devices are properly biased, the midpoint voltage should be equal to one half the supply voltage. Appreciable deviation from this voltage is an indication of some circuit problem. Many amplifiers incorporate bias adjustment controls. The usual method of adjustment is to set the controls to obtain a midpoint voltage equal to one half the supply voltage. Inability to obtain this voltage indicates a circuit malfunction.

4. In direct-coupled circuits, the first transistor can be considered the control for the remaining transistors in the string. A useful testing procedure based on this principle is to cut off the first stage (short emitter to base) while monitoring the midpoint voltage. The midpoint voltage should either drop to zero or rise to the B+ supply voltage, depending on the particular circuit design. (One of the output transistors should saturate while the other cuts off.) If this does not occur, monitor the collector voltage of the first transistor while intermittently cutting off the stage. No voltage change indicates that this stage is defective. A voltage change at this point indicates that the first stage is operating and the problem is either in the driver or output stages. Defective devices in these stages are more easily located by resistance measurements as previously described than by voltage measurements.

5. If the midpoint voltage is either zero or at B+ potential, one of the transistors in the complementary-symmetry pair is either shorted (emitter to collector), or saturated. A saturated transistor can be differentiated from a shorted transistor by shorting the emitter to the base while monitoring the voltage across the transistor. No voltage change indicates a shorted device.

6. If both output transistors short, which is a common occurrence, chances are the emitter resistors will burn out or at least overheat. These are low-value resistors (less than 1 ohm) in high-power amplifiers and must be replaced with the exact value.

117

7. Intermittent noise, hiss, popping sounds, etc, are common problems in audio amplifiers. The best method of tracing down these elusive noise gremlins is through signal-grounding techniques. With the capacitor probe, begin at the input and successively ground the output of each stage. Elimination of the offending noise indicates the defective stage has just been passed. This technique will often pinpoint not only the defective stage but often the very component, such as a coupling capacitor. A capacitor large enough to completely ground the signal without disturbing the dc operation of the stage must be used. In low-impedance circuits, a large value may be required. The 20-μF electrolytic capacitor, as described in Fig. 1-1, is useful for this purpose.

8. The capacitor probe, with a 0.1-μF capacitor, may be used to inject 60-Hz signals into the various stages. Clip the probe to any low-voltage 60-Hz internal source.

9. Most present-day amplifiers are dual channel (stereo). Usually, a malfunction occurs in one channel only unless the problem is related to the power supply. The capacitor probe is, consequently, useful for signal swapping between channels. For example, if one channel is dead while the other channel is normal, a quick determination of whether the problem is in the preamplifier or output stages can be made by jumping the signal output from the preamplifier of one channel into the output stage of the other. Normally, difficult problems such as distortion noise, low volume, etc, can be easily solved by applying this method. An additional feature of having two separate identical amplifiers is that voltage and resistance measurements can be compared.

10. Signal tracing a square-wave signal with the oscilloscope is also a good service procedure. Distortion problems, as well as any oscillations, can be readily detected by this technique. Some amplifiers have a tendency to oscillate at inaudible (ultrasonic) frequencies. The power supply should be checked for adequate high-frequency bypassing in such cases.

Ground the B+ lines at various points with the 0.1-μF capacity probe. Oscillation in high-power amplifiers can sometimes be corrected by connecting a low-value capacitor (0.001 μF) from collector to base of each output transistor.

11. High-gain audio amplifiers are notorious for picking up and detecting rf signals of all kinds. This is especially true if the amplifier incorporates an ac feedback loop from the output stages and is operated with external speakers. The speaker leads act as antennas feeding rf energy back to the driver or predriver stages, where it is detected and amplified. Shielding the speaker leads seldom helps. However, small-value capacitors attached from the speaker leads (at the amplifier) to ground will usually correct the problem. The lowest possible capacitor value needed to correct the problem should be used, since high-frequency response can be deteriorated if too high a value is selected. Collector-to-base feedback as described in Procedure 10 is also effective in some cases. Locate the stage where detection is occurring and use the lowest capacitance value necessary to correct the problem. The offending stage can be readily located by using the signal grounding technique described in Procedure 7.

12. Always operate audio amplifiers with the proper speakers or a dummy load. Do not short across the speaker terminals. In some amplifiers, the output transistors may be damaged if these practices are not followed.

13. When replacing output transistors mounted on heat sinks, be sure to mount them securely and use heat-conductive silicone grease. Many output transistors are electrically insulated from the heat sink. Be certain the insulator or insulators are correctly installed. Check isolation with an ohmmeter before applying power.

14. A loud hum when power is applied usually indicates shorted output transistors. Remove power immediately and check the output transistors with an ohmmeter.

REVIEW

Questions

Q1. A midpoint voltage of other than one half the supply voltage indicates the output transistors are conducting unequally. True or false?

Q2. What will the midpoint voltage in Fig. 6-5 become if C101 shorts? Will the amplifier operate?

Q3. What will the midpoint voltage be in Fig. 6-7 if Q206 shorts from emitter to collector?

Q4. C222 in Fig. 6-7 is shorted. What will the midpoint voltage be when the loudness control is at minimum volume setting?

Q5. In Fig. 6-3, R317 is described as 0.47-ohm resistor. This is probably a typographical error, since a more realistic value would be 47 ohms. True or false?

Q6. If C208 in Fig. 6-1 shorts what will the effect be on the output transistor?

Q7. How will the midpoint voltage be affected if Q302 in Fig. 6-3 shorts from base to emitter?

Answers

A1. False. The output transistors in class-AB, push-pull amplifiers are in a stacked or series configuration. The dc current, therefore, must be the same through both transistors, regardless of the midpoint voltage. A deviation of midpoint voltage from one half the supply does indicate the bias or operating point of the two devices is unequal. The internal resistance from emitter to collector will be unequal, causing unequal voltages to appear across the transistors and a consequent shift in midpoint voltage.

A2. The midpoint voltage will be zero. A shorted C101 removes the forward bias from Q100, cutting it off. Q110, as a result, is cut off. The collector of Q110 falls to ground potential, removing the forward bias from Q120. Whether the amplifier will operate or not depends on the signal level. If the loudness is at maximum, sufficient signal can be applied to the base of Q100 to cause conduction on the positive excursions of the signal (class-C operation). However, the output will be low and distorted.

A3. The midpoint voltage will drop close to zero since the collector of Q206 approaches ground potential, cutting off Q210 and Q214.

A4. The midpoint voltage will be about 34 volts, the supply potential. Forward bias will be removed from Q206, causing the collector voltage to rise and saturating Q210 and Q214.

A5. False. Very low value emitter resistors are employed in high-power audio amplifiers. The value used is critical because it determines the operating point and stability factor of the transistor. Replacements must be of the identical value.

A6. The output transistor will saturate. A shorted C208 removes the forward bias from the driver transistor causing the output transistor to saturate. R207 will overheat since the current will be sufficient to exceed its wattage rating.

A7. The midpoint voltage will not be affected. Transformer coupling isolates the dc operation of the output stage from the previous stages.

CHAPTER 7

FM Stereo Multiplex

Multiplexing refers to the method of encoding a carrier with additional information based on the phase relationship of the additional signal compared to a suppressed carrier. In home-entertainment products, this form of additional modulation conveys the stereo information in fm-stereo radio and the chroma information in color-tv receivers. Phase modulation or phase coding requires a phase-detector circuit of one form or another in the receiver to decipher the multiplexed information. The demodulators employed as decoding circuits in fm-stereo and color-tv receivers are actually phase-comparator circuits.

Stereo broadcasting requires that two separate channels of audio information be transmitted. This could be accomplished in several different ways. For example, the audio information for the right channel could modulate the fm carrier in the conventional manner. The left-channel audio information could be processed in such a manner as to separate it in frequency from the right-channel audio. This can be done by amplitude modulating a separate carrier (sufficiently removed from the highest frequency audio of the right channel) and using the developed sidebands to frequency modulate the fm carrier. At the receiver, two separate detector circuits could then recover the right-channel and left-channel audio signals and apply them to separate audio amplifiers. Such a system would somewhat simplify fm stereo broadcasting, particularly from the transmitter end. However, the problem with a basic system of this type is that it is not compatible with the monaural fm receivers already in existence. Monaural receivers would only be capable of receiving the right channel during stereo transmissions.

Therefore, a compatible system was developed that electrically adds and subtracts the right-channel and left-channel audio signals. In this manner, two separate audio signals are developed. The L + R signal contains all the information contained in the stereo program, but presented in a monaural fashion. An example of L + R information would be the output from a stereo phonograph cartridge with the right and left channels connected in parallel. A monaural receiver detects the L + R information in exactly the same way as when receiving a strictly monaural transmission. The L + R audio signal modulates the fm carrier just as in monaural broadcasting.

The difference or L − R signal amplitude modulates a 38-kHz subcarrier by means of a circuit referred to as a balanced modulator. In such a circuit, the subcarrier is suppressed and two sidebands are created. These sidebands consist of the subcarrier *plus* the modulating frequency on one side of 38 kHz, and the subcarrier *minus* the modulating frequency on the other side of 38 kHz. The two resultant sidebands frequency modulate the main rf carrier of the station. The rate of deviation is essentially the same for both the L + R and L − R signals since they are produced from identical sources. Both contain audio frequencies from 0 to 15 kHz. The two audio signals do not interfere with one another because of the 8-kHz separation between the high end of the L + R sideband and low end of the L − R sideband (Fig. 7-1).

Suppressed-carrier modulation results in better signal-to-noise ratio. The rf energy is divided between the sidebands and the fundamental carrier. Since all of the necessary information is contained in the two sidebands, transmission of the 38-kHz carrier would be inefficient. The L − R sideband signals are phase related to the suppressed 38-kHz carrier. A signal must be produced at the receiver

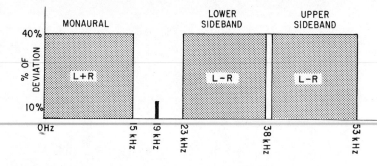
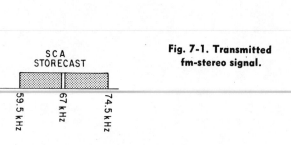

Fig. 7-1. Transmitted fm-stereo signal.

which matches the phase of the suppressed carrier in order to extract the encoded L − R signals. A 19-kHz reference signal, which is of the same phase as the suppressed carrier, is transmitted for this purpose.

Certain stations also broadcast a subscription background-music service for business establishments, which is also a multiplex system. The center or suppressed-carrier frequency in this case is 67 kHz. The sideband energy of this storecast or SCA service extends 7.5 kHz on either side of the suppressed carrier. The fidelity is consequently impaired somewhat as compared to the home-entertainment stereo system. The signal-to-noise ratio is also somewhat deteriorated due to the additional modulation placed on the carrier. Home-entertainment fm-stereo receivers contain trapping circuits to exclude the interfering effects of SCA transmissions.

The means by which the signals are generated at the transmitter is of no importance regarding the understanding or efficient servicing of multiplex receiver circuitry. Because of the low frequencies and simple circuitry involved, multiplex is among the easiest of circuits to troubleshoot. Fig. 7-1 shows the positions of the multiplex signals relative to the allowable bandwidth of the fm station. The precentage of deviation is also indicated.

At the receiver, decoding of the multiplex information can be accomplished in two different ways:

1. Frequency separation and matrixing.
2. Switching or time sharing.

The first of these methods involves separation of the L + R and L − R sideband frequencies. The audio information contained in the L − R sidebands is then detected by reinsertion of the 38-kHz suppressed carrier. The recovered signals are combined with the L + R to form the original right- and left-channel signals. This method is no longer used. It has been made obsolete by the superior and simplified circuitry of the switching type of decoder. A block diagram of a frequency-separation multiplex decoder is illustrated in Fig. 7-2.

Note that the audio signals at the output of the balanced demodulator, L − R and −L + R, represent two possible extremes of audio-signal polarity. These signal polarities, when combined respectively with the L + R signal in the matrix circuit, would produce L only or R only outputs as follows:

$$(L + R) + (L − R) = 2L$$

and,

$$(L + R) + (−L + R) = 2R.$$

It should be understood that the relative phase of the reinserted subcarrier compared to the L − R sideband signals will instantaneously produce signal polarities of infinite variation between the extremes of L − R and −L + R at the demodulator output.

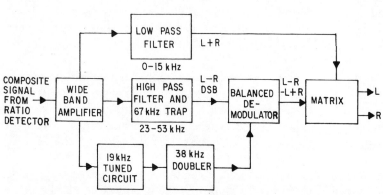

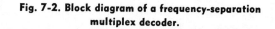

Fig. 7-2. Block diagram of a frequency-separation multiplex decoder.

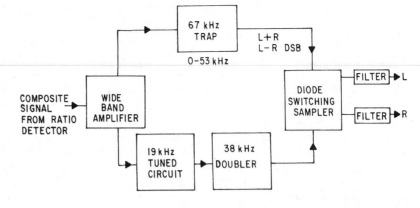

Fig. 7-3. Block diagram of a switching or time-sharing multiplex decoder.

The switching or time-sharing multiplex system of decoding, which is exclusively used at the present time, does not separate the L + R and L − R sideband frequencies. The entire composite signal is switched at the 38-kHz subcarrier rate, extracting bits of left-channel and right-channel audio which are applied to their respective amplifiers after filtering. Fig. 7-3 illustrates a time-sharing multiplex system in block-diagram form. This is the system now used almost exclusively.

FM MULTIPLEX CIRCUITS

Purpose: Decode the composite stereo information into separate left- and right-channel audio signals.

Signals In: Composite stereo information consisting of the L + R signal, L − R suppressed-carrier double-sideband signals, 19-kHz pilot, and any storecast information which may be present.

Signals Out: Separate left-channel and right-channel audio.

Circuit Description

In Fig. 7-4, the composite signal from the output of the ratio detector enters the multiplex circuit at J33. It is applied to a special low-pass filter made up of L26, L27, C26, C27, and C28. All frequencies above 53 kHz are rejected and only those below 53 kHz are fed to Q126, preventing interference from storecast information. This circuit does not use an oscillator or a conventional doubler stage to generate the 38-kHz carrier, but rather, amplifies the 19-kHz pilot from the station and doubles it through the use of two 19-kHz doubler diodes (D26 and D27). These diodes rectify the 19-kHz signal, producing half-wave dc

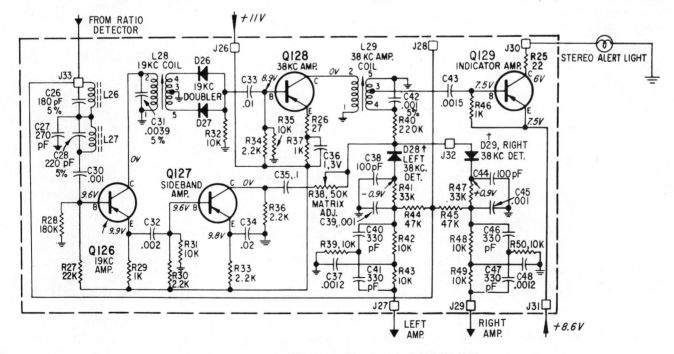

Fig. 7-4. Frequency-separation multiplex decoder circuit.

pulses at a 38-kHz rate. The 38-kHz pulses are then amplified and shaped in the collector circuit of Q128. From the secondary of L29, the 38-kHz signal is applied to an indicator amplifier (Q129) which turns on the stereo alert light, indicating the presence of a stereo signal. The 38-kHz signal is also applied to the matrix circuits through R40. The L − R sideband signal is taken off at the emitter of Q126 and amplified by the sideband amplifier (Q127). From the collector of Q127, the L − R signal is coupled by C35 to the matrix adjustment control (R38). The L − R signal is now combined with the 38-kHz carrier at junction J32. The L + R signal is not passed with the L − R signal because of the high-pass characteristics of the circuits associated with Q126 and Q127.

The L + R information is applied to the matrixing circuit at J28 directly from the storecast trap in the input circuit. Matrixing occurs through algebraic addition and subtraction of the L + R and L − R signals at a 38-kHz rate and the original left- and right-channel information is recovered at terminals J27 and J29 respectively. The RC networks at the output of the detector diodes remove the 38-kHz signal.

In Fig. 7-5, the detected composite fm signal is fed into the multiplex input circuit which is made up of a network of traps consisting of L126, L127, L128, C127, and C128. This trapping complex sharply attenuates signals above 53 kHz, preventing intereference from storecast and adjacent-station information. The input also contains a con-

trol (R126) referred to as an L − R level control. This control, in conjunction with C126, adjusts the phase of the composite signal with respect to the 19-kHz pilot for maximum stereo separation.

The entire composite signal (19-kHz pilot, L + R signal, and L − R sidebands) is fed to the composite amplifier (Q126). Q126 performs two separate functions. First, it acts as an emitter follower to the L + R and L − R signals and applies these signals to the center tap of L131. Second, the 19-kHz component of the composite signal is peaked in the collector circuit of Q126 and again in the base circuit of Q127. This double tuning provides high selectivity and rejects all L + R, L − R, and spurious noise signals.

The 19-kHz pilot signal drives the class-C biased Q127 into conduction and 38 kHz appears at the collector, which is tuned to the second harmonic of 19 kHz by L131 and C135. When a signal is present at the emitter of Q127, Q128 is turned on which biases Q129 into conduction. Q129 acts as a switch, turning on the stereo alert light and indicating the presence of a stereo signal.

The 38-kHz signal, generated by Q127, is inductively coupled through L131 to the switching diodes D125 and D126, which are normally biased on. This alternately biases diodes D125 and D126 off and on in phase with the 38-kHz signal.

Recovery of the original left-and right-channel audio information involves turning on the left and right diodes at precisely the correct phase. When diode D125 is biased on, the 38-kHz carrier is in phase with the L − R sideband and the L − R sig-

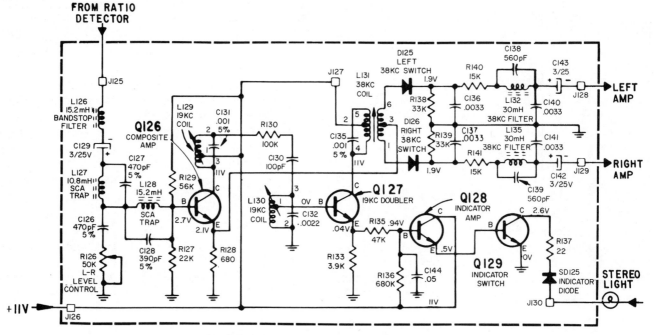

Fig. 7-5. Switching-type multiplex decoder circuit.

nal is detected along with the L + R signal. The R portion of the signal is cancelled through algebraic addition and the left-channel information is recovered at J128. When diode D126 is conducting, the 38-kHz carrier is 180° out of phase with the L − R sideband. Therefore, the −L + R signal is detected along with the L + R signal. The L portion of the signal is cancelled and the right-channel information is recovered at J129.

When the receiver is tuned to a monaural fm station, no 19-kHz pilot appears at the base of Q127, resulting in no 38-kHz subcarrier. With both diodes conducting equally, any signal appearing at center tap of the L131 secondary from the emitter of Q126 will appear equally at both the left- and right-channel outputs (J128 and J129).

The output circuit provides standard fm de-emphasis with component R140, C140, R141, and C141. The 38-kHz detector signal is suppressed by filters made up of L132, C138, L135, and C139. Capacitors C136 and C137 form ac loads for the 38-kHz switching signal. R138 and R139 are the forward-bias dc paths for D125 and D126. C143 and C142 are the output coupling capacitors. Rectifier SD125 provides dc for the indicator light and R137 limits the current.

Test Equipment

1. Jumper lead containing a 0.1-μF capacitor as shown in Fig. 1-1.
2. A vom or vtvm.
3. An oscilloscope.
4. A stereo multiplex generator. A multiplex generator is not usually required to isolate component failures but is useful for alignment purposes.

Service Hints and Procedures

1. Defective multiplex circuitry may cause a complete loss of audio when the function switch is placed in stereo position. A failure of this type is usually a result of an open component in the 67-kHz storecast input trapping circuit. An open emitter-base diode of transistor Q126 (Fig. 7-5) will also cause no audio. Failures of these components can be quickly isolated by connecting the capacitor jumper lead from the output of the fm detector to the input circuitry of the multiplex. Successively bypassing the trapping components and the emitter-base diode of the composite amplifier will isolate the open part by restoring normal operation.

2. No stereo separation is usually a result of some circuit malfunction causing loss of the 38-kHz subcarrier. The first check should, therefore, be for the presence or absence of the 38-kHz signal. The oscilloscope is the ideal instrument to make this check since the 38-kHz subcarrier can be distinguished from the 19-kHz pilot. The defective stage, as a result, is more easily located.

3. Failure of the stereo alert light to illuminate usually indicates a defective indicator transistor, no 19-kHz or 38-kHz signal, improper alignment, or an open indicator bulb. The transistor action can be checked by connecting about a 10k resistor from the collector to the base (Q129 in Fig. 7-4 and Q128 in Fig. 7-5). If the light operates, the transistors can be assumed to be good. If the indicator transistors are good, the light is probably out as a result of no or weak 19-kHz or 38-kHz signals. Check the peak-to-peak waveforms as indicated by the manufacturer with an oscilloscope. Misalignment of the 19-kHz or 38-kHz tuned circuits may cause no light action. These circuits can, in fact, usually be peaked up by monitoring the voltage across the alert light.

4. The manufacturer's instructions should be followed for alignment. The 19-kHz and 38-kHz circuits are easily peaked up with an air signal by observing the particular circuit output with a scope or ac output meter. Unfortunately, most receivers do not produce best separation when the circuits are simply peak aligned. Various techniques for aligning multiplex circuits have been published. However, the simplest and most effective procedure is to use a stereo multiplex generator and follow the manufacturer's instructions.

5. Some receivers employ an adjustable 67-kHz storecast trapping circuit. Storecast interference is commonly referred to as "monkey chatter" and is apparent as a background muttering sound. The 67-kHz trap should be adjusted for minimum interference.

6. The signal-to-noise ratio deteriorates considerably during stereo reception. This becomes apparent as a higher hiss content. Antenna condition and orientation is, for this reason, more critical than that involved with monaural reception.

REVIEW

Questions

Q1. What will be the effect if Q128 in Fig. 7-5 shorts from emitter to collector?

Q2. The receiver in Fig. 7-4 is producing only monaural output. All the dc voltages, transistors, and diodes check okay. What procedure should be followed? Which component is most likely at fault?

Q3. What conclusions should be made if the indicator light operates but no stereo is obtained?

Q4. What conclusions should be made if the indicator light is inoperative and no stereo is obtained?

Q5. What conclusions should be made if the light is inoperative but stereo is obtained?

Q6. What is probably wrong if the indicator light remains lit constantly?

Answers

A1. Q129 will saturate and the stereo indicator light will remain on constantly. Operation will not otherwise be impaired.

A2. A check should be made as to whether the 38-kHz subcarrier is being produced. If this procedure were followed, it would probably determine that the 38-kHz signal is very low in amplitude. Since the dc voltages are correct and the transistors operating, the components most susceptible to failure would be L29 and C36. The coil cannot easily be checked unless it is open. C36 can be easily checked by jumpering it with the capacitor test lead. In this case, normal operation would be restored if the capacitor is open.

A3. In Fig. 7-4 it can be assumed that the 19-kHz and 38-kHz circuits are operating and the problem lies in the detector circuit or the multiplex circuit is improperly aligned. In Fig. 7-5 it can be assumed that 19-kHz circuits are operating and the problem lies in the 38-kHz circuit, the switching circuits, or misalignment.

A4. In Fig. 7-4 a defect probably exists in the 19-kHz or 38-kHz circuits. In Fig. 7-5 the defect is in one of the 19-kHz circuits.

A5. The defect must be an open indicator light or a defective indicator transistor.

A6. The indicator transistor is defective or is biased into saturation as a result of some other defective component. Short the base to the emitter. If the light does not extinguish, the transistor is defective.

CHAPTER 8

Detector Circuits

The term detector or demodulator is a rather broad term applied to a number of circuits used in home-entertainment products. Solid-state detector circuits can be roughly categorized as responding to amplitude, frequency, phase, or any combination of these factors. The multiplex circuitry described in the preceding chapter is actually a combination phase/amplitude detector, but because of the supporting circuitry involved, has been segregated from detectors in general. The active component of the circuits which concern us is usually a semiconductor diode of either germanium, silicon, or selenium construction, depending on application. Because of the relative simplicity of the circuits, troubleshooting generally involves simply checking the front-to-back resistance ratio of the appropriate diode or diodes.

AMPLITUDE DETECTORS

Amplitude detectors or a-m demodulators used in home-entertainment products consist of a single semiconductor diode usually of germanium construction. The detector employed in a-m radio and the video detector in black-and-white and color tv receivers are examples of amplitude detectors.

Purpose: Rectify the amplitude-modulated carrier to obtain either the upper or lower envelope of modulating signal.

Signal In: An amplitude-modulated i-f carrier (Fig. 8-1).

Signal Out:

A-M Radio: (1) Audio signals from about 1 to 5 kHz. (2) A dc voltage, proportional to signal strength, used for automatic gain control purposes.

TV Receivers: Composite video signals from about 0 to 4 MHz. Signals may be of either sync-positive or sync-negative polarity depending on the number of video amplifier stages employed (Fig. 8-2).

Circuit Description

In Fig. 8-3, the amplitude-modulated carrier is rectified by the detector diode producing segments of audio signal at the i-f carrier rate. These segments are filtered by the 0.05-μF capacitor, producing an audio signal without the carrier. The audio level is a function of signal strength. A portion of the audio signal is applied through the 2.2k resistor to a 20-μF filter capacitor. This signal-dependent developed voltage is used as agc to control the bias and gain of the rf and/or first i-f amplifiers.

The operation of the video detector in Fig. 8-4 is essentially the same as the a-m radio detector. Additionally, a video detector incorporates series-tuned and parallel-tuned peaking circuits to bring up the response at the high-frequency end. Video-detector diodes are usually direct coupled to the first video amplifier, which is commonly of emitter-follower configuration to prevent excessive loading of the detector circuit. Some video-detector circuits appear quite complicated since the biasing network of the first video amplifier is a portion of the detector circuit. Peaking and tweet-suppression components also vary in complexity from one manufacturer to another.

Test Equipment

1. A vom or vtvm.
2. An oscilloscope.

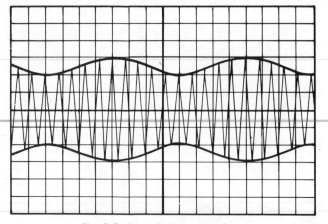

Fig. 8-1. A-m i-f carrier waveform.

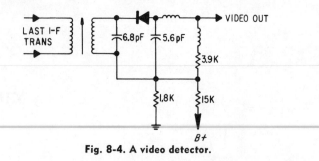

Fig. 8-4. A video detector.

If simple agc is employed, the effect will be a negative picture and poor sync. The agc circuit is also affected in receivers using keyed agc systems.

Service Hints and Procedures

1. The components most susceptible to failure are usually the active devices—in this case, the detector diodes. Check the front-to-back resistance ratio with an ohmmeter. Since the diode is usually of germanium construction, some reverse leakage is normal.

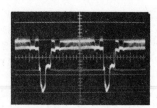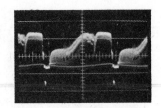

(A) Horizontal rate.   (B) Vertical rate.

Fig. 8-2. Composite video waveform.

2. Observation of the signals into and out of the video detector with an oscilloscope is the best troubleshooting procedure. A detector probe must be used to observe signals prior to detection. Most receivers produce an output from the video detector of about 2 to 3 volts peak to peak.

3. In most receivers the output is at the anode of the video detector, producing composite video with negative-going sync. A reversed video detector will cause a multiplicity of problems.

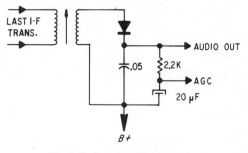

Fig. 8-3. An a-m radio detector.

FREQUENCY DETECTORS

Demodulation of fm requires that the frequency variations be converted to amplitude variations. This could be accomplished simply by a circuit which is tuned so that the center i-f frequency falls at the center of one of the slopes of the resonance curve. To prevent distortion, the frequency swing must be restricted to that portion of the curve which is straight. Since this portion is quite short, the voltage variations developed would be small. This circuit, as a result, is not used. Rather, a sort of push-pull version is employed in the form of either a discriminator or a ratio detector. Present-day fm receivers employ the ratio detector almost exclusively because of its better immunity from amplitude variations.

The discriminator and ratio-detector circuits are practically identical, except for the way in which the diodes are connected and the point at which the audio signal is removed. Both circuits are actually phase-comparator circuits. The frequency variations are converted to two signals of unequal and varying phase. The phase difference is then converted to an audio signal.

Purpose: Convert the frequency variations of the i-f carrier into audio signals.

Signal In: An i-f carrier (10.7 MHz for fm radio and 4.5 MHz for tv) which varies in frequency at an audio rate.

Signals Out: (1) Audio. (2) A positive-going and negative-going dc level used to control the oscillator frequency (afc).

Circuit Description

The primary and secondary of T1 in Fig. 8-5 are tuned to resonate at 10.7 MHz. At resonance, the energy inductively coupled from the primary to the secondary causes the voltage in the secondary to be 90° out of phase from that in the primary. The signal coupled by C1 to the center tap of the T1 secondary remains in phase with the primary. The relative conduction of D1 and D2 will be proportional to the phase difference between these two signals. At resonance, the conduction of the two diodes will be equal. Current will flow in opposite directions, as indicated by the arrows, through load resistors R1 and R2. Since the currents are equal and opposite, the combined voltage drops of R1 and R2 equal zero (point A to ground). As the audio modulation swings the carrier above or below resonance, the energy inductively coupled into the secondary becomes more or less than 90° apart from the primary. Under conditions other than resonance, unequal diode conduction results which produces a voltage from point A to ground. As the signal swings above resonance, this voltage becomes positive in relation to ground. Below resonance, D2 conducts more and the voltage at point A becomes negative with respect to ground. L1 is an rf choke which provides the dc return path for the diodes through R1 and R2. An RC network at the detector output provides proper de-emphasis by attenuating the high-frequency end of the recovered audio signal.

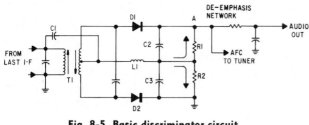

Fig. 8-5. Basic discriminator circuit.

Fig. 8-6 represents a typical ratio-detector circuit. Some variations from this basic circuit may be encountered, depending on the configuration of the ratio-detector transformer T1 and the relation of the signal to ground.

The operation of the ratio-detector circuit is essentially the same as the discriminator except for the way the diodes are connected and the point at which the audio signal is removed. Notice that the diode currents, as depicted by the arrows, are in a series-aiding direction. As in the discriminator detector, the current through R1 and R2 is equal at resonance. The voltage from point A to B will be zero at resonance. As the audio modulation swings the i-f carrier frequency above and below resonance, a negative or positive voltage develops across A and B. The total current through R1 and R2 remains constant and is proportional to signal strength. Only the ratio of current flow between

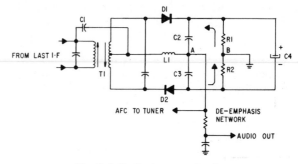

Fig. 8-6. Ratio-detector circuit.

R1 and R2 changes as the signal swings above and below resonance.

The primary advantage of the ratio-detector circuit over the discriminator circuit is a greater immunity to amplitude variations. This immunity is a result of the long time constant of the capacitor C4 across R1 and R2.

Test Equipment

A vtvm.

Service Hints and Procedures

1. Defective diodes are the most common component failures in fm detector circuits. These can be checked by measuring the front-to-back resistance ratio. Disconnect one end of the diode when performing these tests. (Germanium diodes are commonly used and can be expected to have a reverse leakage of several thousand ohms.)
2. Alignment of the ratio-detector or discriminator transformer is a good method of checking circuit operation. The ability to achieve proper alignment usually indicates that circuit operation is proper. Alignment of either the discriminator or ratio-detector transformer is easily accomplished with an on-air signal as follows:
   (a) Place the afc switch in OFF position.
   (b) Carefully tune to an active station.
   (c) Adjust the primary of the transformer for maximum voltage across either R1 or R2 in Fig. 8-5 and across C4 in Fig. 8-6.
   (d) Adjust the secondary of the transformer for zero voltage from point A to ground in both Figs. 8-5 and 8-6.

PHASE DETECTORS

Phase-detector or phase-comparator circuits, as used in home-entertainment products, are of basically two types. However, their operation is quite similar. The first type develops a dc voltage which is proportional to the phase relationship of two separate signals. The dc voltage is, in fact, derived in much the same manner as that produced by a ratio-detector or discriminator circuit. Circuits which fall in this category are: color and horizontal phase detectors and color-killer detectors.

The second type of detector produces a third signal derived from the combination of two signals which vary in phase and amplitude. Color-demodulator circuits and fm multiplex detectors (described in the preceding chapter) are representative of this kind of phase detector.

Horizontal-Phase Detectors

Purpose: Frequency and phase lock the horizontal oscillator.

Signals In: (1) Horizontal sync pulses (Fig. 8-7). (2) A horizontal retrace pulse (Fig. 8-8).

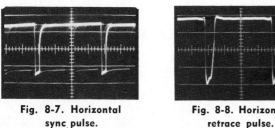

Fig. 8-7. Horizontal sync pulse.

Fig. 8-8. Horizontal retrace pulse.

Signal Out: A dc level proportional to the phase relationship of the input signals.

Circuit Description

The operation of a horizontal phase-detector circuit is essentially the same as an fm discriminator circuit. In Fig. 8-9, the incoming sync signals are applied through the differentiating network (R526 and C523) to the common cathodes of D520. This develops equal but opposite voltages across

R521 and R522. These opposing voltages cancel each other and no dc voltage will be developed. However, simultaneously with the application of the sync pulses, a positive-going pulse from the horizontal output transformer is applied through isolating capacitor C524 to the phase detector. This pulse is shaped to a sawtooth form by the RC network (C129 and R129). Now, if the sync pulse occurs as the sawtooth voltage is passing through its zero point, there will be no output from the phase detector. However, if the oscillator is running faster than the sync, the output will be positive; if it is slower, the output will be negative.

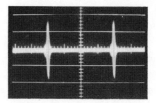
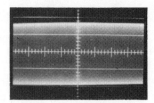

Fig. 8-10. Burst signal at horizontal sweep rate.

Fig. 8-11. 3.58-MHz oscillator signal at horizontal sweep rate.

The dc output is applied to a reactance circuit, which shifts the oscillator frequency an amount proportional to the degree of phase mismatch between the two signals. The time constant of components R519 and C517 prevents the circuit from *hunting* or oscillating. An open component will create a condition commonly referred to as "pie-crusting", which is a periodic displacement of groups of horizontal scan lines. This results in zig-zag effect to any vertical edges in the picture.

Color Phase Detectors

Purpose: Lock the frequency and phase of the color oscillator.

Signals In: (1) A 3.58-MHz burst signal occurring at a horizontal rate (Fig. 8-10).

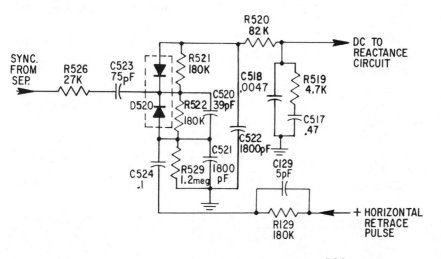

Fig. 8-9. Horizontal phase detector.

(2) A continuous 3.58-MHz signal from the color oscillator (Fig. 8-11).

Signals Out: A dc level proportional to the phase relationship of the input signals.

Circuit Description

The operation of a color phase-detector circuit is essentially the same as an fm ratio-detector circuit. In Fig. 8-12, burst signals of equal amplitude but opposite phase are applied to the two diodes. When the bottom of the burst-transformer secondary winding is negative and the top positive, the two diodes conduct equally and capacitors C737 and C738 charge equally, but with opposite polarity. The voltage from the junction of the two diodes to the center tap of the transformer (ground) will be zero. At this time, no current will flow between these points. During the next half cycle of burst the top of the burst-transformer winding becomes negative and the bottom positive, reverse biasing the diodes. Capacitors C737 and C738 now discharge through R737A and R737B, the matched 470k resistors. The voltage difference between the junction of the two resistors and the center top of the burst transformer (ground) will also be zero at this time.

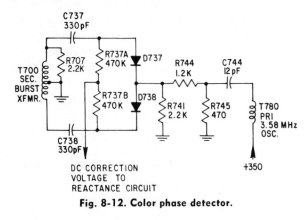

Fig. 8-12. Color phase detector.

A sample of the 3.58-MHz oscillator signal is applied to the junction of the two diodes. If the 3.58-MHz oscillator signal is passing through zero (in sync) when the burst is at peak amplitude, the voltage at the junction of the diodes or resistors will remain zero relative to ground and no correction voltage is developed. If the 3.58-MHz oscillator signal is not passing through zero at this time, the diodes do not conduct equally, capacitors C737 and C738 do not charge or discharge equally, and a voltage differential is developed between the junction of the resistors and ground. This correction voltage controls the reactance tube which maintains the oscillator at the same frequency and phase as the burst signal.

Color-Killer and ACC Detector

Purpose: (1) Produce a dc level which controls the gain of one or more chroma amplifiers. (2) A dc level which turns the last chroma amplifier on when burst is present and off when burst is not present.

Signals In: (1) A 3.58-MHz burst signal occurring at a horizontal rate (Fig. 8-10). (2) A continuous 3.58-MHz signal from the oscillator (Fig. 8-11).

Signal Out: A dc level proportional to the amplitude of the burst signal.

Circuit Description

All color receivers incorporate some form of color-killer circuit which turns off one of the chroma amplifiers during black-and-white reception. Most modern receivers also employ some sort of automatic color gain control. Both of these functions can be controlled by the burst signal. The circuit described here detects both the presence and the amplitude of burst signal, thereby developing a dc level to operate the color-killer and automatic chroma control circuits. The circuit illustrated in Fig. 8-13 is basically the same as the color phase detector just described. The primary difference is the takeoff point for the 3.58-MHz oscillator signal. In the phase detector, the signal is taken from the primary of T780, while in this circuit the signal is taken from the secondary. The phase relationship between burst and the 3.58-MHz oscillator signal is, therefore, different in the two circuits. The color phase detector is strictly a phase-detector circuit. The killer detector, on the other hand, is partially a peak- or amplitude-detector circuit.

The phase-detector circuit is a balanced detector producing zero output when the burst and 3.58-MHz oscillator signals are properly phased. The killer-detector circuit is always unbalanced with D736 conducting more than D735 and, therefore, producing a positive output. This biases on Q740 which functions as a dc amplifier. The biasing path for Q740 is up through R747, R735B, D736, R739, and R740 to B+.

The combined effect of the burst signal and the oscillator signal is such that D735 conducts an amount proportional to burst amplitude, causing the detector output to the base of Q740 to become less positive and reducing its conduction. The reduced voltage across the emitter resistor (R747) is used to control the gain of the first chroma amplifier and the switching mode of the color-killer circuit.

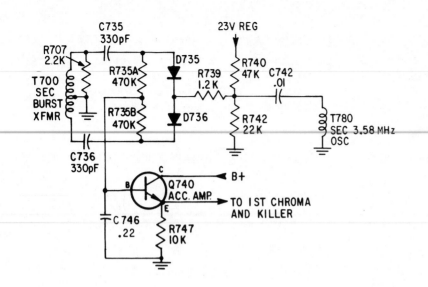

Fig. 8-13. Color-killer and acc detector circuit.

Test Equipment

1. A vtvm.
2. An oscilloscope.

Service Hints and Procedures

1. Defective diodes are the most likely component failure. Check the front-to-back resistance ratio with an ohmmeter. Disconnect one end of the diode when making this test. Germanium diodes are generally used in color-killer and color-phase detector circuits and can be expected to have several thousand ohms reverse leakage. Horizontal phase-detector diodes are commonly molded into a single three-lead package with the center lead common to both diodes. The diodes are of a special selenium construction which results in very low reverse leakage. Almost no leakage should be measurable with most ordinary vtvms or voms. The forward resistance should be about 500 ohms.

2. Most phase-detector circuits produce zero output voltage when the oscillator which it controls is phase locked. If either input signal is lost, the output voltage deviates from zero by a small amount, usually about 1 volt. This causes the oscillator to operate slightly off frequency. An out-of-sync condition will be apparent on the crt either as "barber poling" for a color phase detector or out-of-sync video for a horizontal phase detector. In either case, five or six diagonal bars should be observed. Many bars out of sync indicate the correction voltage has deviated from zero by considerable amount (six or seven volts is common). When the oscillator is this far off frequency, a defective phase-detector diode is the most common malfunction. Substitution is usually the best method of checking horizontal phase-detector diodes.

3. Signal tracing the input signals with an oscilloscope is the most effective shop service procedure.

REVIEW

Questions

Q1. What is the purpose of the 0.05-μF capacitor in Fig. 8-3?

Q2. Will the video output from the circuit in Fig. 8-4 contain positive- or negative-going sync pulses?

Q3. What peak-to-peak voltage can be expected at the output of circuit in Fig. 8-4?

Q4. Which is more commonly used for fm detection—the ratio detector or discriminator? Why?

Q5. The output from a ratio detector contains both an ac and dc component. Of what use is the dc component?

Q6. Is L1 in Fig. 8-6 inductively coupled to T1?

Q7. What is the purpose of C4 in Fig. 8-6?

Q8. The secondary of a ratio detector transformer is adjusted to produce zero dc output voltage when properly tuned to an active station. True or false?

Q9. What symptoms will be noticed if C517 in Fig. 8-9 opens?

Q10. What would be the effect of a shorted T700 primary in Fig. 8-13?

Q11. What will the emitter voltage of Q740 in Fig. 8-13 be if D736 opens?

Answers

A1. This capacitor filters the rf carrier, producing a clean audio signal.

A2. Since the output is from the anode, the sync will be negative going.

A3. Most receivers produce 2 to 3 volts peak to peak of composite video at the detector.

A4. The ratio detector because of its inherent immunity to audio modulation (noise).

A5. The dc component provides afc action. This voltage causes the capacity of a varactor diode in the tuner to change, swinging the oscillator frequency one way or the other.

A6. No. L1 is an rf choke.

A7. Provides a-m immunity. The long time constant of C4 across R1 and R2 makes the circuit insensitive to amplitude changes such as noise.

A8. True.

A9. Pie-crusting. (The picture will display a zig-zag pattern.)

A10. The color will be out of sync due to the absence of the burst signal.

A11. Zero volts. The biasing path for Q740 will be open.

COLOR DEMODULATORS

The chroma information is multiplexed onto the rf carrier in much the same manner as described in the chapter on fm stereo multiplex. The suppressed carrier in this case is 3.58 MHz which must be regenerated at the receiver to demodulate the color information. This is accomplished by a local 3.58-MHz oscillator which is frequency and phase locked by the transmitted burst signal (the equivalent of the 19-kHz pilot carrier in fm multiplex). The regenerated subcarrier and color sideband signals are compared in the demodulator circuit. The instantaneous phase and amplitude relationship of the two signals develops a third signal which represents one of the three color signals.

Color demodulators are of two basic types. One type is the so-called X and Z demodulator which develops the red and blue color-difference signals only. The green signal is derived in the color-difference amplifiers (see the chapter on chroma amplifiers). The other type of demodulator develops all three color-difference signals (a block diagram of this system was shown in Fig. 5-5).

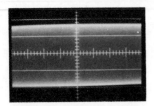

Fig. 8-14. 3.58-MHz chroma signal at horizontal sweep rate.

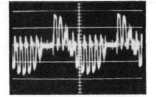

Fig. 8-15. Keyed color-bar signal at horizontal sweep rate.

The color-difference signal which is developed by a particular demodulator is dependent on the phase relationship of the redeveloped subcarrier to the color sideband signals. For example, the red and blue demodulators are identical except that the 3.58-MHz subcarrier applied to one is about 90° out of phase with the 3.58-MHz subcarrier that is applied to the other.

Color demodulators are switching circuits and, consequently, the active devices may be transistors or diodes. In either case, color-demodulator circuits are slightly modified versions of the basic ratio-detector circuit previously described. The signal levels, however, are such that a switching or sampling action occurs just as in fm stereo-multiplex circuitry.

Purpose: Provide two (or three) signals of varying amplitude which are used to drive the control grids of the crt.

Signals In: (1) Chroma sideband signals of approximately 3.1 MHz to 4.1 MHz (3.58-MHz center frequency suppressed). This signal (Fig. 8-14) is applied equally to each demodulator. (2) The 3.58-MHz oscillator signal is supplied to the X demodulator. This signal is phase shifted about 90° from burst by the 3.58-MHz output transformer to achieve demodulation on the X axis and recover the R − Y signal. (3) The 3.58-MHz oscillator signal is shifted an additional 90° and applied to the Z demodulator to achieve demodulation on the Z axis and recover the B − Y signal. (4) If G − Y is demodulated, the 3.58-MHz oscillator signal is shifted an additional 120° and applied to the G − Y demodulator.

Signals Out: Amplitude variations at frequencies of from about 0 to 0.5 MHz (Fig. 8-15).

Circuit Description

Transistors Q750 and Q751 in Fig. 8-16 are reverse biased by the positive voltage applied to the emitters. This voltage, approximately 1 volt, is developed by divider resistors R792, R748, R793, and R752. The transistors are gated into conduction during negative excursions of the 3.58-MHz oscillator signal. The 3.58-MHz signal applied to the emitter of the Z demodulator is phase shifted about 90° from the signal on the X demodulator emitter. This phase shift is accomplished by LC network L753 and C753.

The chroma signal is applied equally to the base of each demodulator from the third chroma ampli-

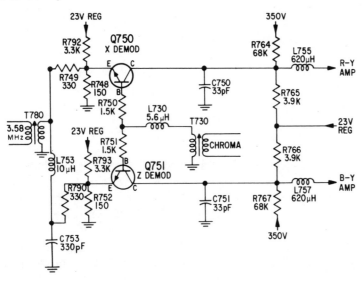

Fig. 8-16. X and Z demodulators employing transistors.

fier. The output signal amplitude from each demodulator depends on the amplitude and, consequently, the phase of the chroma signal at its base during the time the transistor is switched into conduction by the 3.58-MHz oscillator reference signal. This output signal is, therefore, proportional to the peaks of the original I and Q signals produced at the transmitter. (I and Q signals are proportional to but shifted in phase from the X and Z signals presently detected in most receivers.) The original I and Q signals developed at the transmitter are 90° out of phase. Since the reference 3.58-MHz signal applied to the demodulators is also 90° out of phase, signals proportional to the original I and Q color signals are recovered as amplitude variations in the collector circuits of the demodulators. The signal outputs of the X and Z demodulators, when properly combined, produce the three color color-difference signals which control the red, blue, and green crt guns. (See color-difference amplifiers.)

Notice that the collector supplies are derived from two voltage sources, a 23-volt supply and a 350-volt supply. This produces higher collector voltages which allows a greater signal swing than if only the 23-volt supply were used. LC networks L757, C751, L755, and C750 filter out the 3.58-MHz subcarrier. L730 prevents radiation of high-frequency oscillations.

The operation of the X and Z demodulators in Fig. 8-17 is the same as Fig. 8-16 except for the phase angle of the applied 3.58-MHz oscillator signal. This phase difference is produced by a network consisting of C417, R433, and L406 in the input circuit to the X demodulator. This phase difference is approximately 90°, allowing the demodulators to produce the $R - Y$ and $B - Y$ color signals which are 90° apart in phase relationship.

In operation, the Z demodulator has the 3.58-MHz signal applied to D401 and D402 through C416. The positive and negative excursions of this signal alternately forward and reverse bias these diodes. The chroma signals applied through C419 to D401, and through C421 to D402 will cause one or the other of these diodes to conduct, depending on the amplitude and phase relationship of the chroma signal to the 3.58-MHz signal. This process results in a series of pulses at a 3.58-MHz rate, with amplitude variations corresponding to the amplitude of the $B - Y$ appearing at the junction of R425 and R427. The amplitude variations of the $B - Y$ signal appear across R436 for amplification by the Z amplifier. C423 and C425 act as 3.58-MHz filters producing clean amplitude variations of color information.

The X demodulator operates in a like manner, with the $R - Y$ signal variations appearing across R441 at the base of the X amplifier. Neither de-

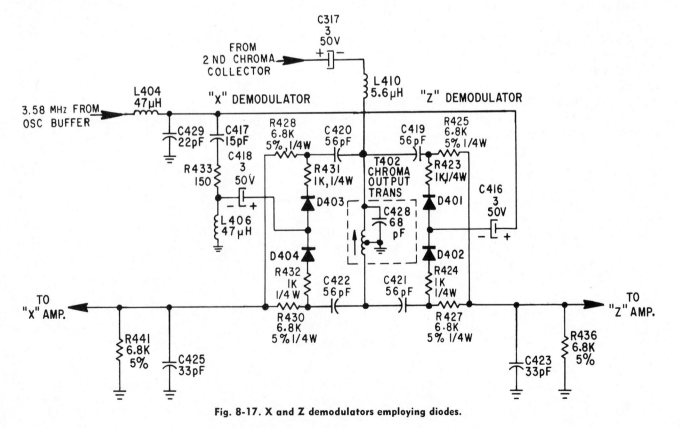

Fig. 8-17. X and Z demodulators employing diodes.

modulator operates during the horizontal retrace period when burst is present in the chroma signal, since the output of the 3.58-MHz buffer amplifier is interrupted during this time.

The dc bias conditions of the detector diodes are established by the long time constant of C416 and C418. The diodes are signal biased, since the charge retained on the capacitors tracks with the signal and determines the peak-to-peak level required for conduction. This action is analogous to signal biasing of sync-separator or burst-amplifier stages.

Test Equipment

1. A capacitor jumper lead containing a 0.1-μF capacitor.
2. A vtvm or vom.
3. A color-bar generator (optional since most color problems can be located using color broadcast signals).

Service Hints and Procedures

1. Demodulator operation can be checked by injection of a 60-Hz signal. Clip the capacitor probe to a low-voltage, 60-Hz source and inject the signal in place of the chroma signal. If the demodulator is operating, a hum bar should appear across the crt face. The X demodulator should produce a red hum bar and the Z demodulator should produce a blue hum bar.
2. Voltages and front-to-back resistance ratios of diodes and transistors can be checked with a vom or vtvm. As with most circuits, the active device (transistor or diode) is most likely to fail.
3. Signal tracing the input and output signals is the best service procedure in the shop. A station signal or the output from a color-bar generator work equally well since either can be readily observed with most modern scopes.

REVIEW

Questions

Q1. Since color demodulators are switching circuits, the signal from the 3.58-MHz reference oscillator is much greater in amplitude than the chroma signal. True or false?

Q2. Are the transistors in Fig. 8-16 operating class-A, B, or C?

Q3. What is the purpose of C750, C751, L755, and L757 in Fig. 8-16?

Answers

A1. True. The chroma sideband information is sampled at a 3.58-MHz rate. The 3.58-MHz reference oscillator can be considered the switch driving signal and is normally at least ten times the amplitude of the chroma signal.

A2. Class-C. Without a signal, the transistors are reverse biased. The oscillator signal must rise above this level before conduction occurs.

A3. These components are 3.58-MHz filters. Excessive 3.58-MHz signal at the crt grids causes a fine checkered pattern to appear across the entire picture.

CHAPTER 9

Control and Timing Circuits

Many circuits in home-entertainment products, particularly in tv receivers, can be loosely described as control and timing circuits. Most of the circuits in this category are quite simple and straightforward, varying little from one manufacturer to another. Examples are the sync separator, burst amplifier or separator, 3.58-MHz oscillator, etc. Some circuits, such as agc systems, vary considerably from one manufacturer to another. However, the general theory of operation remains about the same. Categorization of circuits in this manner is somewhat arbitrary, since the detector circuits described in the preceding chapter could also be considered control circuits. The intent is to adhere to the original format of grouping circuits by function as closely as possible.

AGC SYSTEMS

Automatic gain control systems applied to tv receivers may be either of two types: simple or keyed. Simple agc systems employ the dc component resulting from the rectification of the video signal by the video-detector diode. This dc voltage is proportional to signal strength (and video content) and is used to control the gain of the rf and i-f amplifiers. Both a-m and fm radio receivers accomplish gain control in this manner.

Simple agc systems have several disadvantages, such as poor noise immunity and the tendency of the gain to change with video content. However, keyed agc systems operate only during sync time and, as such, are insensitive to video variations. Transistor amplifiers do not offer quite the range or linearity of gain control common to vacuum-tube amplifiers. Because of this and other inherent shortcomings, simple agc systems are seldom used in solid-state tv receivers.

Either the keyed or simple agc circuits can be designed to develop a dc control voltage which increases the forward bias as the signal becomes stronger (forward agc). They can also be designed to develop a control voltage which decreases the forward bias as the signal increases (reverse agc). Forward agc has become the more popular of the two methods, particularly when bipolar devices are used, because of better overload immunity. Transistors designed for forward agc operation have also become more readily available.

Any amplifier can be designed to provide a gain reduction by reducing the forward bias (reverse agc action). The primary gain-reducing mechanism is an increase of the intrinsic emitter-to-base impedance with reduced conduction. (See the section on gain in Part 1.)

Forward agc systems are only applicable to high-frequency amplifiers. For this reason, rf and i-f amplifiers in a-m radios always utilize reverse agc systems. Several factors are involved in reducing gain by increasing the forward bias. The major consideration, however, is a decrease in high-frequency beta which is due to an increase in the emitter-to-base capacitance. The increased capacitance loads the input circuit and also increases the transit time, thereby reducing high-frequency gain. Depending on circuit design, two other factors are also involved. These are the collector-to-base capacitance (involving Miller effect) and the loading effect of the transistor on the tuned output circuit. Miller effect is significant only if the circuit contains a collector load resistor large enough to cause appreciable change in collector voltage with increased forward bias. The voltage between the collector and the base will decrease with increased bias. This causes capacity to increase, loading the input circuit and reducing gain.

In a properly designed circuit, the loading effect of the transistor on the output tank circuit should not be significant. Normally, gain reduction in this manner is undesirable since the Q factor of the tuned circuit is reduced. The loading effect is minimized by connecting a swamping resistor across the tuned circuit which is lower in value than the impedance change of the emitter-collector circuit of the transistor.

Some transistors are specifically designed for operation as forward-agc, high-frequency amplifiers. Greatest gain will be realized at a specific value of emitter current, usually several milliamperes. Deviation from this value in either direction reduces the gain as shown by the gain curve in Fig. 9-1.

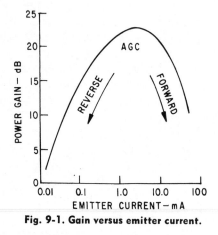

Fig. 9-1. Gain versus emitter current.

TV AGC Circuits

Purpose: Develop dc voltages proportional to signal strength which control the gain of the rf and i-f amplifiers.

Signal In: (1) Composite video (Fig. 9-2).
  (2) Horizontal retrace pulse (Fig. 9-3).

Signal Out: Dc voltage proportional to signal strength.

Circuit Description

The circuit shown in Fig. 9-4 is an agc system which develops a reverse agc voltage. This agc circuit controls the gain of a tube-type rf amplifier and FET i-f amplifiers. The conduction level of

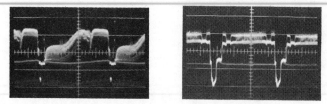

(A) *Vertical sweep rate.*   (B) *Horizontal sweep rate.*

Fig. 9-2. Composite video signal.

Q570 is established by the setting of R156 and the amplitude of the sync pulses applied to the base. During normal operation, the service switch is in the 23-volt position producing proper agc operation at some setting of R156, the agc control. When the service switch is in midposition, the emitter of

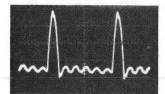

Fig. 9-3. Horizontal retrace pulse.

Q570 is grounded through R159 causing the transistor to saturate. The saturation of Q570 develops maximum negative agc voltage which cuts off the tuner and the first and second i-f amplifiers, producing a blank raster to facilitate purity adjustments.

Conduction of Q570 occurs only during horizontal retrace time when a positive-going pulse from

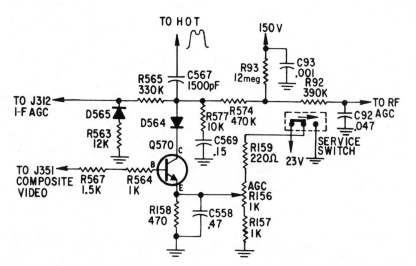

Fig. 9-4. Reverse keyed-agc circuit.

the horizontal output transformer is applied to the collector. The degree of conduction, which is proportional to the amplitude of the sync pulse present on the base, determines the level of the charge on C567. The greater the signal level, the more Q570 conducts and the higher the negative voltage becomes at the anode of D564. D564 prevents the charge on C567 from forward biasing the collector-base junction of Q570 and leaking off between retrace pulses.

R93 provides the rf delay, consisting of a small positive voltage applied to the rf amplifier, which allows the amplifier to operate at maximum gain until a signal level of approximately 1000 microvolts is received. At this time, the increasing negative voltage present at the anode of D564 overcomes this positive voltage and begins reducing the gain of the rf amplifier. The negative voltage increases in proportion to signal strength and reaches about −5 volts with the signal level at approximately 100,000 microvolts.

The i-f agc is established at about 4 volts positive under no-signal conditions by the voltage divider resistors in the i-f circuit. Under these conditions, the i-f amplifiers are operating at maximum gain. As the signal increases, the negative

voltage developed at the anode of D564 reduces this positive voltage in proportion to signal strength. The i-f agc voltage will vary from about +4 volts to −0.6 volts, depending on signal strength.

D565 is reverse biased under all but very strong signal conditions. Signal conditions strong enough to drive the i-f agc voltage negative will cause D565 to conduct, switching R563 into the circuit. R563 and R565 then become a voltage divider which greatly reduces the amount of negative agc excursion possible. The purpose of this circuit is to improve tracking between the rf and i-f agc circuits under very strong signal conditions (100,000 microvolts or greater).

Fig. 9-5 illustrates an agc circuit which develops both forward and reverse agc voltages. This agc system controls a tube-type rf amplifier which requires reverse agc and transistor i-f amplifiers utilizing forward agc.

The agc-keyer transistor Q470 develops a negative voltage at the collector which is proportional to signal strength. This voltage is applied to the base of the agc amplifier transistor Q460, controlling its conduction. As the conduction of Q460 changes, the amount of forward (positive) bias

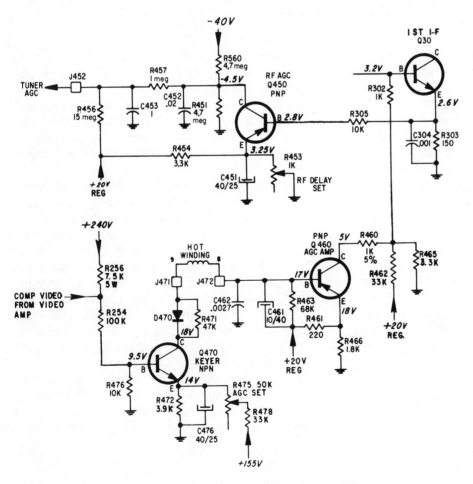

Fig. 9-5. Keyed-agc circuit providing forward i-f agc voltage and reverse rf agc voltage.

applied to the first i-f amplifier varies, controlling the gain of this stage.

The average conduction of Q470 is established by the base voltage developed through divider resistors R476, R254, and R256 and the emitter voltage determined by the setting of R475. A positive-going composite video signal is also applied to the base of the keyer transistor. A positive pulse from the horizontal output transformer is applied to the collector of Q470 and occurs at the same instant that a horizontal sync pulse appears on the base of this transistor. The conduction of Q470, which takes place only during the horizontal retrace interval, charges C462 in proportion to the amplitude of the sync pulse appearing at the base of this transistor. The negative charge on C462 is applied to the base of the agc amplifier Q460.

The conduction path of Q470 is through the emitter-base junction of Q460. As the conduction of Q470 increases with signal, a greater negative charge is developed on C462 causing Q460 to conduct more. Since Q460 is a pnp transistor, the collector voltage becomes more positive with increased conduction. This rising positive voltage applied to the base of Q30, the first i-f amplifier, increases its conduction causing a reduction in gain (forward agc). Increased conduction of Q30 causes the positive voltage across the emitter resistor R303 to rise. When the amplitude of the sync pulse applied to the base of the agc keyer Q470 is reduced, a reverse action takes place and the voltage across R303 decreases. It is this voltage that controls the rf agc circuit.

The conduction of the rf agc transistor Q450 is controlled by the emitter voltage established by R453 and the base voltage developed across R303. Increased conduction of Q30 causes a rising positive voltage across R303 which decreases the conduction of Q450. Decreased conduction of Q450 causes the collector voltage to become more negative, reducing the gain of the rf amplifier (reverse agc action). A positive voltage is also applied to the rf amplifier through R456 to provide rf agc delay. This allows the rf amplifier to operate at maximum gain under very weak signal conditions. As signal strength increases, the negative voltage developed by Q450 overcomes the positive voltage and reduces the rf gain.

Test Equipment

1. A jumper lead.
2. A vtvm.
3. An oscilloscope.
4. A bias supply.

Service Hints and Procedures

1. If agc problems are suspected, clamp the i-f and rf agc voltages as called out on the schematic. Restored operation indicates that the problem is in the agc circuit, and not an i-f or tuner problem. A variable dc supply or bias box is best for this purpose. However, the voltage available at the ohmmeter terminals of a vom can also be useful for test purposes. (See Chapter 1.)

2. The operation of the agc system can be easily checked by alternately cutting off and saturating the agc keyer transistor. Shorting the emitter of the agc keyer to the base, should cut off the transistor. No agc voltage should be developed, causing the picture to be overloaded and out of sync. If this does not occur, the keyer transistor is probably defective. Saturate the transistor by connecting a 10k resistor from B+ (for an npn) or ground (for a pnp) to the base. The transistor should saturate, developing maximum agc voltage and cutting off the rf and i-f amplifiers. There should be no picture or an extremely weak picture. Again, if this does not occur, the keyer transistor is likely to be defective.

3. A saturated keyer transistor can be simulated by shorting the emitter to the collector. If picture cutoff does not occur when this is done, the problem is a shorted keyer diode or is external to the keyer stage.

4. Defective (usually leaky or shorted) keyer diodes are a common problem in keyed agc circuits. A defect of this sort results in little or no agc voltage being developed and an overloaded, out-of-sync picture.

5. Often, agc filter capacitors are a source of trouble, producing many different symptoms ranging from hum in the picture to various oscillating conditions. Bridge the rf and i-f agc lines with a low-value electrolytic capacitor (about 1 or 2 μF). When replacing defective capacitors in any of the agc output circuits, be certain to use the correct value. These capacitors serve more than just a filter function. The time constant of the entire agc system is important to provide good noise immunity and to prevent airplane flutter, etc.

6. Check input waveforms with an oscilloscope. No agc voltages can be developed if the horizontal pulse from the flyback is missing or the composite video signal is not present at the base.

REVIEW

Questions

Q1. What is the function of the keyer diode?

Q2. What would occur if C476 in Fig. 9-5 becomes shorted?

Q3. What would occur if the horizontal output transformer winding shown in Fig. 9-5 were to open?

Q4. Why is R93 in Fig. 9-4 often referred to as an rf delay?

Q5. What will be the picture condition if the service switch in Fig. 9-4 is activated?

Answers

A1. The keyer diode prevents loss of the developed agc voltage by preventing the collector-base junctions of the keyer transistor from becoming forward biased between retrace pulses.

A2. Q470 will saturate developing maximum agc voltage and ultimately cutting off the rf and first i-f amplifiers. A blank raster will result.

A3. Neither Q470 nor Q460 will conduct. No agc voltage will be developed and an overloaded, out-of-sync picture will result.

A4. The application of a positive voltage to the rf amplifier through R93 delays the gain reduction of this stage until a specific signal level is attained; that is, until sufficient agc voltage is developed to overcome this positive voltage and begin gain reduction. Usually from 500 to 1000 microvolts of signal is required before rf gain reduction takes place.

A5. Q570 will saturate causing the developed agc voltage to be maximum. The i-f and rf amplifiers will cutoff, resulting in a blank raster. This particular feature aids in color-temperature and purity setup adjustments.

AFC SYSTEMS

The higher-priced versions of most modern color-tv receivers feature circuitry which automatically corrects for mistuning by the customer. Although the systems used vary from one manufacturer to another, they are essentially ratio-detector or discriminator circuits that develop a correction voltage proportional to the deviation in frequency from the picture carrier (45.75 MHz). This correction voltage is applied to a varactor diode (voltage-variable capacitor) in the tuner which brings the oscillator back on frequency. The system functions similar to the afc circuits used in fm radio receivers. Most tv receivers employ an integrated circuit which performs this function with a minimum of external circuitry.

Purpose: Provide automatic fine tuning to correct for improper tuning by the viewer.

Signals In: Intermediate frequencies from the third video i-f amplifier.

Signal Out: A differential dc level proportional to the frequency deviation from a 45.75-MHz center frequency.

Circuit Description

Integrated circuit IC360 in Fig. 9-6 is actually four circuits built into one device. These are: a zener-diode voltage regulator, a broad-band amplifier, a discriminator detector, and a dc amplifier. The regulator portion of the IC between pins 8 and 10 maintains a normal voltage of 11 volts. The input circuit to the broadband amplifier is between pins 7 and 6. The external reactive components in the input circuit peak up the high side of the response curve (toward 47.25 MHz). The amplifier output at pin 2 is externally inductively coupled to the detector circuit at pins 1 and 3. T360 and L362 are tuned to a 45.75-MHz center frequency (picture carrier). The detector circuit is internally connected to the dc amplifier, which produces a differential voltage output proportional to the degree of frequency deviation from 45.75 MHz. When the uhf and vhf fine tuning is correct, zero voltage exists between pins 4 and 5. When the receiver is mistuned, the differential voltage developed between pins 4 and 5 is applied to the varactor diodes in the uhf and vhf oscillators to bring them back on frequency.

The afc defeat switch disables the afc circuit by shorting the two output points together and producing zero differential voltage. A 100-μF nonpolarized capacitor is connected across the differential output. This capacitor produces a delay effect, when switching the afc circuit on, to prevent abrupt changes in afc action.

Test Equipment

1. A vom.
2. Sweep and marker generators.

Service Hints and Procedures

1. Most of the active circuitry is contained within the IC. Therefore, troubleshooting involves measuring the voltages around the IC and comparing them with those on the manufacturer's schematic. The manufacturer's instructions should also be followed for alignment purposes.
2. Alignment of afc circuits is quite difficult if not impossible with an air signal. Do not bother the afc adjustments unless you have the proper equipment and are prepared to perform an alignment.

SYNC SEPARATORS

Purpose: Separate the sync pulses from the composite video signal and apply these pulses to the horizontal phase-detector and vertical integrator circuits.

Signal In: Positive-going composite video (Fig. 9-7).

Signal Out: Negative-going sync pulses (Fig. 9-8).

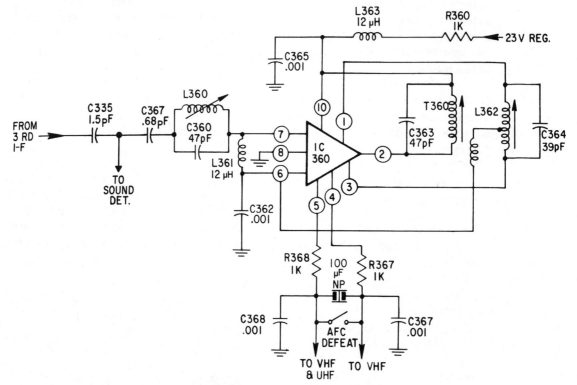

Fig. 9-6. An afc circuit utilizing an IC.

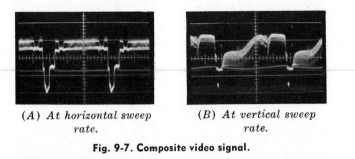

| (A) At horizontal sweep rate. | (B) At vertical sweep rate. |

Fig. 9-7. Composite video signal.

Circuit Description

In Fig. 9-9, the composite video signal appearing across R352 is applied to the base of Q550, the sync-separator transistor. The positive-going composite video signal is of sufficient amplitude to drive Q550 into a switching mode. The emitter-base diode produces signal rectification causing

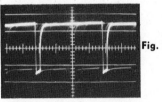

Fig. 9-8. Sync pulses at horizontal sweep rate.

the base to go negative relative to ground. The time constants of the R and C components in the base circuit are such that the bias produced by a combination of R552 and signal rectification will hold the transistor in cutoff mode on all but the positive sync pulse tips. The collector circuit signal will, therefore, consist of only negative-going, horizontal and vertical sync pulses. The collector voltage is derived from divider network R150, R551, and R550, which produces about 50 volts.

Test Equipment

1. A vtvm or vom.
2. An oscilloscope.

Service Hints and Procedures

1. Analysis of the input and output signals with an oscilloscope is the best service technique.

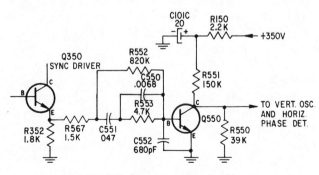

Fig. 9-9. Sync separator circuit.

Most sync-separator circuits produce a high-amplitude signal. The circuit illustrated in Fig. 9-9 provides sync pulses with amplitudes of about 50 volts peak to peak.

2. The output-meter function of the vom can be used in lieu of a scope to check the output signal at the separator. About 10 volts ac will be measured in the example illustrated when the circuit is functioning normally.

3. The presence of video in the sync signal indicates a failure of one of the components in the base circuit of the sync separator. These RC components develop the bias which results from signal rectification.

BURST AMPLIFIERS

Purpose: Provide phase and frequency sync to the 3.58-MHz oscillator. A 3.58-MHz phase reference is, thereby, established for color demodulation.

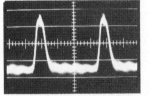 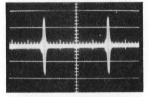

| Fig. 9-10. Horizontal keying pulse with burst signal. | Fig. 9-11. Output from burst amplifier. |

Signals In: (1) A positive-going horizontal keying pulse (Fig. 9-10). This retrace pulse is delayed slightly to occur at the time burst is present. (2) Composite chroma.

Signal Out: A 3.58-MHz sine wave occurring at the horizontal retrace interval (Fig. 9-11).

Circuit Description

In Fig. 9-12, the positive-going horizontal retrace pulse is delayed slightly by integrating components C718, R718, and R703. This is done so that the retrace pulse will occur at the time when burst is present, thereby keying the transistor into conduction. Reverse bias is maintained during scan time by the discharge time constant of C700 and R701. An amplified 3.58-MHz signal, occurring during the horizontal retrace interval, appears in the collector circuit. This signal is applied to the killer and oscillator phase-detector circuits. Comparison of the phase of the burst signal and the oscillator output develops a dc voltage which is used to control the oscillator frequency and phase. Comparison of burst and oscillator amplitudes in the killer detector develops a dc voltage

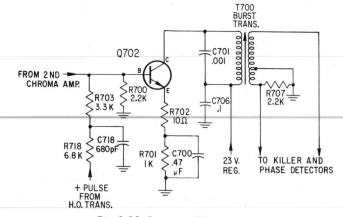

Fig. 9-12. Burst amplifier circuit.

used to control the operation of the color-killer and acc circuits.

Test Equipment

1. A vom or vtvm.
2. An oscilloscope.

Service Hints and Procedures

1. Analysis of the input and output signals with an oscilloscope in the best service technique.
2. An inoperative burst amplifier will cause an out-of-sync color display (barber poling).
3. An inoperative burst amplifier will also be apparent when attempting an afpc (automatic frequency and phase control) adjustment. The afpc adjustments usually involve measuring a dc voltage at some designated test point in the phase-detector circuit and peaking the burst transformer for maximum output. This adjustment will, of course, have no effect if the burst amplifier is inoperative.
4. The presence of chroma information in the burst signal indicates an improper bias level or a defective transistor. An open emitter-bypass capacitor (C700 in Fig. 9-12) can cause this type of problem.

3.58-MHz REFERENCE OSCILLATOR

Color reference oscillators are of basically two types: the so-called "brute-force" variety and the type which is controlled by a phase-detector circuit. The "brute-force" oscillator is frequency and phase controlled by direct application of the burst signal. The phase-detector controlled oscillator employs a varactor diode which swings the oscillator phase in accordance with the dc output voltage of the phase detector. Servicing of either circuit involves about the same techniques. The afpc adjustment procedure will vary somewhat between

the two types and the manufacturers instructions should be followed.

Purpose: Develop a 3.58-MHz reference signal which duplicates the frequency and phase of the original suppressed subcarrier at the transmitter. This signal is required to demodulate the color signals.

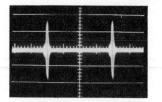
Fig. 9-13. Burst signal at horizontal sweep rate.

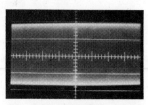
Fig. 9-14. 3.58-MHz signal from oscillator.

Signals In: (1) A dc level proportional to the phase relationship of burst signal to the 3.58-MHz oscillator (phase-detector controlled). (2) Burst occurring at a horizontal sync rate (Fig. 9-13) in "brute-force" type oscillators.

Signal Out: A 3.58-MHz sine wave, frequency and phase locked to burst (Fig. 9-14).

Circuit Description

Fig. 9-15 illustrates a modified version of a basic oscillator. The 3.58-MHz crystal, the varactor diode, and the 470-pF and 100-pF capacitors form the tank circuit. The voltage across the varactor diode is adjustable manually and also controlled automatically by the correction voltage from the phase-detector circuit. The capacity of the varactor diode is inversely proportional to the reverse bias voltage applied. Since the diode is part of the tank circuit, the phase of the oscillator can be shifted and locked by the correction voltage developed in the phase-detector circuit.

In Fig. 9-16, a 3.58 MHz crystal Y401 is the main frequency-determining component. Capacitors C404 and C405 and the secondary winding of T401, in conjunction with the burst signal, are employed to keep the oscillator on frequency and in phase with the burst signal.

The sine-wave output appearing across the collector load resistor R411 is coupled to the oscillator buffer amplifier through C410. An additional output is taken from the base of Q402. The amplitude of this signal increases in proportion to the amplitude of the burst signal applied to the oscillator. It is utilized to control the acc circuit which, in turn, regulates the gain of the chroma amplifier.

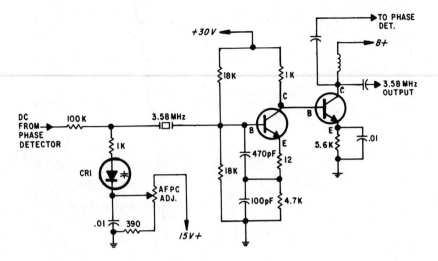

Fig. 9-15. Phase-detector controlled 3.58-MHz oscillator.

R407 varies the 3.58-MHz oscillator amplitude and is adjusted to produce proper acc action.

Test Equipment

1. A vom or vtvm.
2. An oscilloscope.

Service Hints and Procedures

1. An inoperative 3.58-MHz oscillator may produce different symptoms from one brand of receiver to another. Certain receivers lose color completely, others develop a distinctive color such as green or magenta. The type of demodulators and color amplifiers used by a particular receiver determines the characteristics of the symptoms displayed.
2. The oscillator output can be viewed directly with the oscilloscope, which is the best means of determining oscillator operation.
3. Most afpc alignment instructions include peaking the oscillator for maximum output. Usually this involves reading a dc voltage developed by

detection of the 3.58-MHz signal. As a result, a dead oscillator is quickly discovered when this adjustment is attempted.
4. A dead oscillator is usually caused by a defective transistor, the 3.58-MHz crystal, or an open coil.

NOISE GATES

Most television receivers incorporate some type of noise-gate circuitry. The primary intent of such circuits is to prevent the sweep oscillators (particularly vertical) from interpreting noise pulses as sync pulses. Noise-gate circuitry, thereby, minimizes annoying vertical flip and jitter. Noise-gate circuitry varies considerably from one set to another. However, the circuit illustrated in Fig. 9-17 is quite representative regarding basic operation.

Purpose: To prevent extraneous noise pulses of greater amplitude than the sync pulses

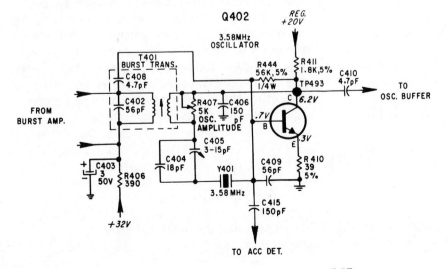

Fig. 9-16. "Brute-force" 3.58-MHz oscillator.

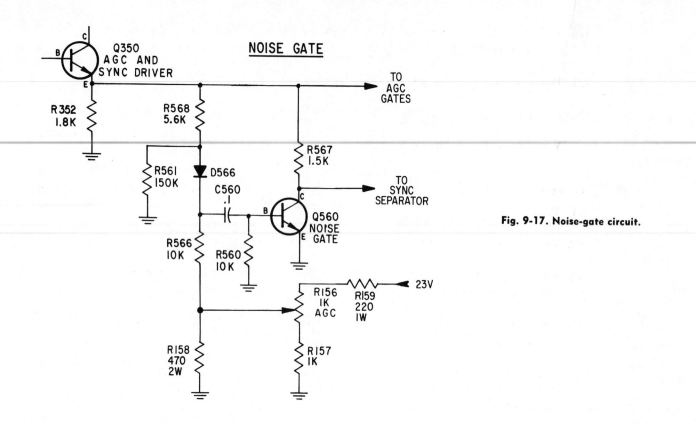

Fig. 9-17. Noise-gate circuit.

from upsetting the sync separator or agc circuits.

Signals In: (1) Positive-going pulses of greater amplitude than the sync pulses are applied to the base (noise pulses). (2) Positive-going composite video is applied to the collector.

Signal Out: Inverted noise pulses at the collector which cause cancellation.

Circuit Description

The base and emitter of Q560 are normally at ground potential, causing Q560 to be reverse biased or cut off. Diode D566 is reverse biased to a degree corresponding to the setting of the agc control. Should a noise occur which is greater in amplitude than the positive-going sync pulse, diode D566 will conduct, charging C560 and forward biasing Q560 for the duration of the pulse. The inverted pulse appearing at the collector will cancel the positive noise pulse appearing in the composite video signal, which is also present at the collector.

Test Equipment

A vtvm or vom.

Service Hints and Procedures

1. Most signal conditions are such that the noise-gate transistor can be entirely removed from the circuit, or disabled by shorting the base to emitter, with no degradation of receiver performance. A turned-on noise gate removes the composite video signal from the agc and sync-separator circuits. Therefore, when any sync or agc problem is encountered, the first step should be to disable the noise-gate transistor.

2. The noise-gate circuit operation can be evaluated by artificially creating noisy reception. An old electric razor, or any brush-type motor with the spark-suppression capacitors removed, generates noise interference when operated near the antenna. Performance can be evaluated by operating the noise generator with and without the noise-gate circuit functioning.

REVIEW

Questions

Q1. The afc circuit illustrated in Fig. 9-6 is essentially the same type as used in fm radio receivers. True or false?

Q2. Will the base voltage of Q550 in Fig. 9-9 change from signal to no-signal operation?

Q3. What occurs if C700 in Fig. 9-12 opens?

Q4. Is CR1 in Fig. 9-15 forward or reverse biased?

Q5. What symptoms will be apparent if C560 in Fig. 9-17 shorts?

Answers

A1. True.

A2. Yes, the base will go negative due to signal rectification.

A3. The color will be out of sync. The burst will be of insufficient amplitude and not separated properly from the chroma signal.

A4. Reverse biased. Varactor or variable-capacity diodes are not normally operated in a forward-biased mode.

A5. The picture will appear overloaded and out of sync. Q560 becomes forward biased and shunts the composite video signal for the agc and sync circuits to ground.

CHAPTER 10

Sweep Circuits

Solid-state sweep circuits vary considerably from one manufacturer to another. RCA, for example, currently employ SCRs (silicon controlled rectifiers) as switches to develop the horizontal deflection for their solid-state color receivers. Motorola incorporates two horizontal output transistors operating in parallel to produce horizontal scan. Some solid-state receivers (color and black and white) develop high voltage with a vacuum-tube high-voltage rectifier, while others utilize a solid-state rectifier or tripler circuit. The present trend in horizontal deflection circuit design seems to be toward a single bipolar output transistor. Solid-state devices with higher voltage and current ratings are continually being developed, providing better performance at reduced cost.

Many color receivers presently on the market are of hybrid construction, the low-power circuits employing transistors and the other circuits, particularly the sweep circuits, utilizing vacuum tubes. The sweep circuits chosen as representative and described in this chapter are used in a 19″ black-and-white receiver. The same basic principles apply to color receivers. Color sweep circuitry differs from black-and-white sweep circuitry primarily in regard to the power needed to accomplish sweep and develop the required high voltage. Additional circuitry which develops the signals necessary for convergence and pincushion correction are also common to color receivers. These circuits are usually quite trouble free and specific convergence and pincushion adjustment instructions are covered in detail by the manufacturer.

VERTICAL SWEEP SYSTEMS

The transistor vertical sweep circuits operate in much the same manner as tube-type vertical sweep systems. The familiar sawtooth waveforms are present in both circuits and are usually generated in the same manner. Transistor vertical sweep systems usually have an additional stage between the oscillator and output stage. The other areas of difference are, of course, the relatively low impedance, low voltage, and high current of solid-state devices compared with vacuum tubes.

Vertical Integrator

At the output of the sync separator, frequency-selective networks are used to distinguish between the horizontal and vertical sync pulses. The horizontal frequency control circuit is capacitively coupled directly to the output of the sync-separator circuit. Since the vertical sync pulse also contains horizontal sync information, the horizontal sync circuits can accept all the information from the sync separator. The vertical oscillator circuit is coupled to the sync separator by a frequency-selective network designed to remove the horizontal sync pulses from the vertical sync pulses. This is a simple RC time constant circuit referred to as a vertical integrator. As shown in Fig. 10-1, the RC values are such that the relatively short duration horizontal sync pulses do not develop appreciable voltage across the network. However, the longer duration horizontal sync pulses which make up the vertical sync interval do develop significant output voltage.

Vertical Oscillator

A popular type of vertical oscillator is the blocking oscillator. Strictly speaking, a blocking oscillator does not produce the required sawtooth waveform, but rather produces a pulse. When a relatively large capacitance is placed across the output, the pulse does become the required sawtooth

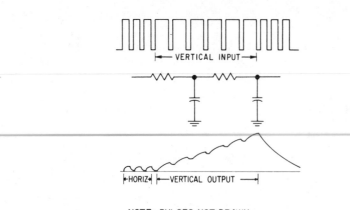

NOTE: PULSES NOT DRAWN TO SCALE.

Fig. 10-1. Vertical integrator circuit.

waveform. The blocking oscillator in Fig. 10-2 is biased in such a manner that the incoming sync pulse from the vertical integrator circuit triggers the transistor into conduction at the proper time. The conduction of the transistor provides a low-impedance path through the transistor and blocking transformer. The capacitors across the circuit discharge through the transformer and transistor, inducing a voltage in the primary which drives the transistor harder and harder into conduction until saturation is reached. This period of conduction produces the retrace portion of the sawtooth waveform. During saturation of the transistor, the base circuit components charge to a negative potential that causes the device to turn off, thus allowing capacitors in the collector circuit to recharge at an exponential rate. This action applies a sawtooth voltage to the vertical driver transistor.

The time constants of the components in the collector and base circuits determine the frequency of oscillation. A control is placed in the base circuit

to vary the time constant of these components and, therefore, functions as a vertical hold control. The base circuit also contains a diode which shunts the transformer primary. This diode clips the negative voltage pulse which occurs during retrace time to a safe amplitude, preventing possible breakdown of the transistor base-emitter junction.

Vertical Driver

The output of the oscillator is directly coupled to the vertical-driver stage as shown in Fig. 10-2. The emitter-follower configuration is used as the driver stage to properly match the low input impedance of the output transistor.

The vertical size and linearity controls are located in the driver stage. The linearity control is part of a feedback network from the emitter to the base of the driver transistor which causes the slope of the trace pulse to change. The emitter resistor of the driver stage functions as a gain or vertical height control. Vertical control circuitry of this type is very stable and the controls are noninteracting. That is, the linearity control affects only linearity and the height control only size. Neither control affects the frequency of the oscillator.

Vertical-Output Stage

In Fig. 10-2, the driver stage is directly coupled to the vertical-output transistor. The output stage is forward-biased to produce class-A operation with the application of a positive-going sawtooth voltage. The output at the collector is inductively coupled to the vertical deflection coils by means of an autotransformer. The vertical-output stage, as well as the horizontal-output stage, should not be

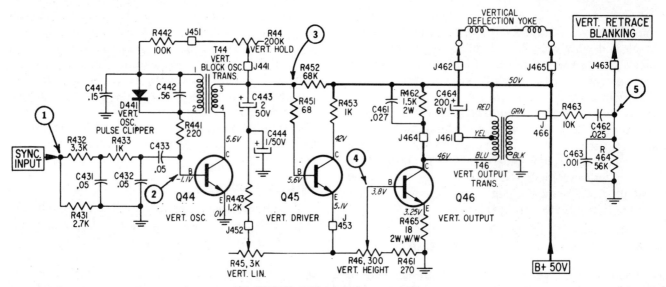

Fig. 10-2. Typical vertical sweep circuit.

150

operated with the yoke disconnected. During retrace time, a voltage pulse of sufficient amplitude to destroy the transistor could develop across the transformer.

The vertical-output transformer also has a secondary winding which is used to produce negative-going pulses during retrace time. These pulses are applied to the control grid of the crt to cause cutoff during vertical retrace, thereby, blanking out vertical retrace lines.

Purpose: Develop vertical deflection yoke currents which properly deflect the crt electron beam.

Signals In and Out: See waveforms in Fig. 10-3.

Circuit Description

In Fig. 10-2, positive-going sync pulses are coupled to the base of the oscillator transistor Q44 through C433. Vertical integration is accomplished by R432, R433, C431, and C432. A slight amount of forward bias is produced by R441, R442, and

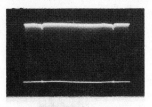

18V P-P
30~

(A) Input to vertical integrator.

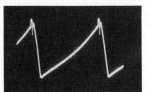

4.5V P-P
30~

(B) Base of vertical blocking oscillator.

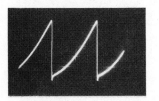

12V P-P
30~

(C) Output of vertical oscillator.

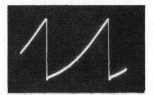

7V P-P
30~

(D) Base of vertical output transistor.

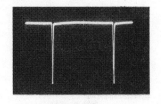

(E) To yoke.

90V P-P
30~

Fig. 10-3. Vertical sweep waveforms.

R44. However, during operation, the base is negative with respect to the emitter due to rectification of the oscillations being produced. C441 and C442, in conjunction with R442 and the vertical hold control R44, form a time-constant network to control the oscillator frequency. Diode D441 clamps any negative swing after retrace time, protecting the transistor from the high-voltage pulse developed across the primary of T44.

Capacitors C443 and C444 discharge through the blocking transformer when Q44 is conducting. R452, the collector load resistor, determines the charge time of the capacitors. The charge-discharge time of these capacitors partially determines the frequency of oscillation. Two capacitors are used in series to provide a feedback path from the emitter to the base of the driver transistor. This feedback path is the linearity-control circuit made up of components R45, R443, C443, and R451. The positive-going sawtooth voltage from the oscillator is direct coupled to the driver transistor Q45. The resistor R452 also provides forward bias to produce class-A operation.

The driver stage is an emitter-follower configuration. R453 is the collector load resistor, while R46 and R461 are the emitter resistors. R46 is also the vertical height control (gain control) which direct couples the positive-going sawtooth voltage to the class-A operated output transistor Q46. R465, the emitter resistor for Q46, is not bypassed and provides degeneration and stabilization in this stage. The collector load is the vertical-output transformer that inductively couples the sweep voltage to the deflection yoke by autotransformer action. R462 and C461 form a damping network across the transformer to reduce the amplitude of the retrace or kickback pulse, preventing damage to the transistor. C464 couples the output transformer to the yoke to prevent the dc component from causing centering problems. Components R463, R464, C462, and C463 form a shaping network for the retrace pulse which is applied to the control grid of the crt for blanking purposes.

Test Equipment

1. A vom or vtvm.
2. An oscilloscope.

Service Hints and Procedures

1. Solid-state vertical deflection circuitry is usually direct coupled from the oscillator to the output stage. The transistors behave as a single device and failure of any one disturbs the operating point of all the others. As a result, signal injection techniques such as hum injection (a

common procedure when servicing tube-type circuitry) is not usually a successful method of isolating a defective stage.

2. A shorted or saturated vertical-output transistor will usually cause a fuse or circuit breaker to open. This is also true of the horizontal-output transistor. Excessive current drain is usually a result of a defect in either the vertical or horizontal sweep circuitry. The transistors in most receivers are mounted on, but electrically insulated from, a metal heat sink. The electrical connection to the collector is usually through the mounting screws. Therefore, the output stage can be disabled in most cases simply by removing the mounting screws. This procedure will help isolate the cause of excessive current drain.

3. Shorted or open transistors are the most commonly encountered problems. An ohmmeter check of front-to-back resistance ratio of all the transistors in the circuit is the quickest and most effective service procedure. The transistors usually can be checked in circuit if the R × 1 scale is used. (See Chapter 1.) Some circuits which do not employ a blocking capacitor in the collector circuit require that the collector be disconnected in order to check the output transistor. This is due to the very low resistance of the vertical-output choke or transformer.

4. When replacing output transistors be certain to mount securely, use heat-conducting silicone grease, and install the proper insulators. Also bear in mind that since the transistor is insulated from the chassis, the case often is "hot" and constitutes a shock hazard.

5. Observing the various waveforms with an oscilloscope is the most effective service technique in cases other than complete failure. The waveforms for a typical vertical sweep circuit are illustrated in Fig. 10-3.

6. Do not operate the receiver with the yoke disconnected. In some circuits a pulse can be developed across the vertical-output transformer which may damage the output transistor.

7. Do not disable transistors in sweep circuits by shorting the base to the emitter. Damage to the output transistor may result.

8. Output transistors with low beta are a common cause of vertical instability and/or linearity distortion. The low gain causes inadequate scan which can, in many instances, be compensated for by improper adjustment of the height and linearity controls. It is these improper settings which create the unstable conditions.

9. Operation on low line voltage should be avoided. Damage to the output transistor may result if the oscillator fails to start.

HORIZONTAL SWEEP SYSTEMS

Solid-state horizontal sweep circuits are usually made up of three or four stages, depending on the type of oscillator employed. The frequency and phase is controlled by a conventional phase-detector circuit which may, or may not, be followed

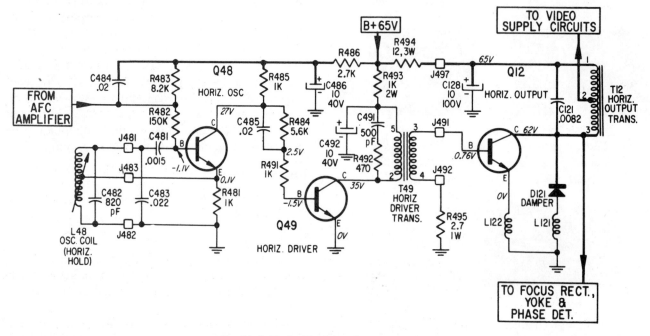

Fig. 10-4. Typical horizontal sweep circuit.

by a dc-amplifier stage. The three-stage deflection system illustrated in Fig. 10-4 consists of an oscillator, a driver, and an output stage. Some of the general characteristics of each stage will be described here.

Horizontal Oscillator

Solid-state horizontal oscillators can, as in the case of tube-type oscillators, be of various circuit configurations. The primary difference between the two is that the transistor output stage must be driven with a square-wave voltage while a tube-type output stage is driven with a modified sawtooth voltage. The reason for this difference will be explained subsequently.

Requirements of the horizontal oscillators are:

1. Good noise immunity.
2. Good frequency stability.
3. Sufficient output to eliminate the need for additional stages of amplification.
4. Capable of being frequency controlled by application of a dc voltage to the base of the transistor.

Several oscillator types meet these requirements to various degrees. The three most-common types are the blocking oscillators, the multivibrator, and the sine-wave oscillator. Although each type has certain advantages, the sine-wave oscillator has the best noise immunity and a relatively high output.

The oscillator shown in Fig. 10-4 is of the modified sine-wave type. Normally, a sine-wave oscillator cannot be frequency controlled by application of a dc voltage, due to the long time constant of the components in the base circuit. This can be overcome by shortening the time constants in the control circuit, allowing the bias point to shift with the application of a dc voltage from the phase detector.

Horizontal Driver

The horizontal-driver (or buffer) stage performs three basic functions. (1) It provides an amplified pulse to properly drive the output transistor. (2) It shapes the pulse from the oscillator into the required square wave. (3) In effect, it is a tuned circuit which determines the pulse width and, together with the resonance of the output circuit, determines flyback or retrace time.

Both the driver and output transistors are operated as pulse amplifiers or switches. That is to say, the transistors have only two modes of operation, completely turned on (saturation) or completely cutoff. As previously mentioned, a tube-type horizontal-output stage is driven by a modified sawtooth voltage, while transistor output stages are driven by a square-wave pulse. This difference stems from the fact that transistors, being low-impedance devices, are inherently excellent switches. Operating a transistor in a switching mode has the distinct advantage of requiring very little power dissipation from the transistor. This becomes obvious when one considers that $P = I^2R$. Since R is very low when the transistor is in saturation, the actual power to be dissipated within the device occurs primarily during switching time. For a square wave, this time is very short. The superiority of transistors as switches results in sweep systems which outperform tube-type sweep circuits in certain areas. Improved linearity is one direct result of this factor.

The driver transistor produces a square-wave pulse output as a result of the manner in which the base is driven. The signal from the horizontal oscillator, a modified sine wave, is passed through a shaping network before being applied to the driver base. Although this applied signal is not a true square wave, the negative excursion of this signal is great enough to drive the transistor sharply into cutoff. The resulting collector output is essentially a square wave, since the transistor is either completely on or completely off.

Horizontal-Output Stage

The purpose of the output stage remains the same for transistor circuits as for tube circuits; to provide beam-scanning currents through the yoke and to produce a high-voltage pulse for rectification and use as the second-anode voltage for the crt.

The development of the proper scan currents requires an analysis of the basic functioning of a transistor horizontal-output circuit. This is most easily described by referring to a simplified representation of a typical transistor horizontal-output stage and observing the ac currents flowing in the circuit. Refer to Fig. 10-5, where:

L = combined inductances of the yoke and output transformer,
C = combined capacitance of the windings and capacitive components across the windings, and
D = damper diode.

Understanding the operation of this basic circuit requires that the signal applied to the base of the transistor be compared to the current through L (see Fig. 10-6). The current through L, which includes the deflection yoke, determines the posi-

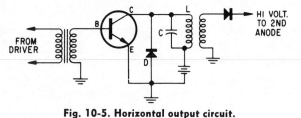

Fig. 10-5. Horizontal output circuit.

tion of the electron beam. The relationship of the voltage appearing across the resonant circuit is also significant.

During time A to B, the transistor is biased on (saturated). Current flows through L, increasing at a linear rate determined by the inductance of L and the voltage applied. This sweeps the beam from the center to the right edge of the crt. At time B, the voltage on the base goes sharply negative, cutting the transistor off. In fact, the base-emitter diode conducts in the reverse direction much like a zener diode, assuring complete cutoff. The transistor cutoff causes the LC tank circuit to oscillate.

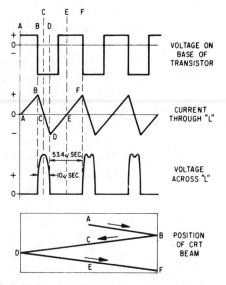

Fig. 10-6. Voltage and current waveforms for circuit in Fig. 10-5.

The time period from B to D represents one-half of this oscillation. The frequency of the oscillation (about 71 kHz) is determined by the resonance of the LC combination. This half cycle of oscillation is the retrace time, during which the beam is rapidly moved across the screen from right to left. The actual mechanism of this retrace function, step by step, is as follows:

1. During transistor conduction from time A to B, current flows through inductor L, building a magnetic field around the coil. At

this time, the right half of forward scan is accomplished.

2. The transistor is cut off at time B, the magnetic field around the coil collapses, and the LC tank circuit begins the first cycle of oscillation. At time C, the current through L is zero and the voltage across L is maximum. Capacitor C is now fully charged. This is the B to C portion of the retrace time.

3. Capacitor C will now discharge in the opposite direction through the coil L. This action produces the C to D portion of the retrace time.

4. At time D, the oscillation is stopped when the damper diode becomes forward biased. The D to E portion of forward scan is completed as a result of the diminishing current through the coil L as the damper is conducting. The damper diode also improves horizontal linearity and reduces the power dissipation of the output transistor.

5. At time E, the current through the damper diode and coil L has dropped to zero, thus completing a scan cycle. The transistor is then biased on to begin conduction and the next sweep cycle.

Purpose: Provide beam-scanning currents through the horizontal deflection coils. Produce a high-voltage pulse which, when rectified, develops the second-anode voltage.

Signals In and Out: See waveforms in Fig. 10-7.

Circuit Description

In Fig. 10-4, bias is applied to the base of the oscillator transistor Q48 by R482 and R483. The actual measured bias, when operating, will be negative with respect to the emitter due to signal rectification by the base-emitter diode. The dc correction voltage is applied at the junction of the two base bias resistors. C484 filters any remaining 15,750-Hz signal, ensuring a pure dc correction voltage. R481 is the emitter stabilizing resistor. In-phase feedback from emitter to base is accomplished by a tap on the oscillator coil L48 and capacitor C481. The frequency of oscillation is determined by the resonance of the LC circuit consisting of L48, C482, and C483. The oscillator coil also functions as the horizontal hold control. R485 is the collector load resistor.

The output of the oscillator is direct coupled to the base of the horizontal driver transistor Q49 through a waveshaping network consisting of C485, R484, and R491. This network modifies the

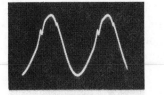

17V P-P
7875∿

(A) *Emitter of horizontal oscillator.*

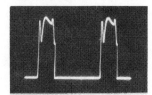

22V P-P
7875∿

(B) *Output of horizontal oscillator.*

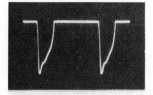

12.5V P-P
7875∿

(C) *Base of horizontal driver.*

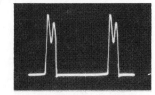

106V P-P
7875∿

(D) *Output of horizontal driver.*

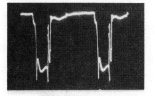

7V P-P
7875∿

(E) *Base of horizontal output transistor.*

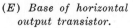

525V P-P
7875∿

(F) *Collector of horizontal output transistor.*

Fig. 10-7. Horizontal sweep waveforms.

pulse wave to more closely approximate the required waveform.

The driver transistor Q49 is forward biased by R484 and R491. Again, signal rectification causes the base to be driven negative with respect to the emitter during operation. The signal on the base drives the transistor in a switching mode, either to complete cutoff or saturation. The collector load is provided by the primary of the driver transformer. The transformer primary is tuned by R492 and C491. C492 provides a low-impedance ac signal path to ground and also reduces the amplitude of the retrace pulse appearing across the transistor. Capacitor C486 and resistor R486 form a B+ filter network for isolation between stages.

The base of the output transistor Q12 is driven by the signal from the horizontal driver transformer T49. R495 is a current limiting resistor to avoid overload of the base-emitter junction. The coil L122 in series with the emitter, and L121 in

series with the damper diode, are suppression inductances to prevent parasitic and flyback transformer oscillations from entering other circuits. Capacitor C121 tunes the output transformer to produce proper retrace time. R493 and C128 constitute a B+ filter network for isolation between stages.

Various other voltages and signals are developed by the horizontal-output stage as follows:

1. The retrace pulse is stepped up by transformer action and rectified by a tube or solid-state rectifier to produce high voltage for the second anode.
2. Focus voltage is either developed from an additional winding and rectifier, or is stepped down from the second-anode potential by a resistive divider network. Many current color receivers employ the latter method, utilizing high-voltage resistors encapsulated in an epoxy package.
3. Convergence and pincushion-correction pulses are developed by the horizontal output transformer.
4. A portion of the energy developed by the horizontal sweep circuit is commonly rectified and filtered, providing boosted B+ for circuits requiring a higher potential than that produced by the power supply. For example: the video output transistor.
5. Retrace blanking pulses.
6. Agc keying pulses.
7. Burst-amplifier keying pulses.
8. Phase-detector comparison pulses.

Test Equipment

1. A vom or vtvm.
2. An oscilloscope.

Service Hints and Procedures

1. A shorted or saturated horizontal-output transistor will cause the fuse or circuit breaker in the receiver to open. The stage can usually be disabled by removing the transistor mounting screws as explained under Service Hints and Procedures in the Vertical Sweep section.
2. Shorted or open transistors are the most commonly encountered problems. An ohmmeter check of the front-to-back resistance ratio of all the transistor diode junctions in the sweep circuit is often the most effective procedure. The first two stages are usually direct coupled and act as a single device. Use the R × 1 scale of the ohmmeter. The input and output impedances associated with the output transistor are

very low. In most cases, it will probably be necessary to remove, or at least disconnect, either the base or emitter lead when checking the front-to-back diode ratios of the output transistors.

3. Tracing the various waveforms with an oscilloscope is the most effective service technique in cases other than complete failure. Typical waveforms are illustrated in Fig. 10-7. An advantage of solid-state sweep circuitry is that all the waveforms may be safely observed (because of the relatively low voltages) with a conventional service oscilloscope. Output transistors will overheat and fail if driven improperly. The driving pulse must be a square wave with sharp leading and trailing edges to assure operation as a switch. Bear in mind, however, that the signal observed at the base during operation is a result of both the driving signal and the reflection of the retrace voltage pulse. Often it is desirable to observe the driving pulse only. This can be accomplished by disconnecting the collector (remove the mounting screws) of the output transistor. A dummy load may also be used as described in Procedure 6.

4. Loss of drive signal in the circuit illustrated merely causes the output transistor to cut off. This is true of most solid-state sweep circuitry.

5. All transistors in the horizontal sweep circuits are usually operating class-C, as switches. During operation, the transistors appear to be reverse biased due to signal rectification. Therefore, measurement of the base voltage is often a quick means of determining if a particular stage is operating.

6. Do not operate the receiver without the yoke connected. Unloading the horizontal-output transformer can cause an increase in the amplitude of the retrace pulse, damaging the output transistor. The receiver should also not be operated with the output transistor removed. Unloading of the driver transformer may damage the driver transistor. A power diode may be used as a dummy load to replace the emitter-base circuit of the output transistor.

7. When replacing output transistors be certain to mount securely, use heat conducting silicone grease, and install the proper insulators. Remember that the case of the output transistor usually has a potential which is above chassis ground.

8. Check the damper diode before replacing a defective output transistor. Although an open damper diode may produce no observable symptoms, long-term overdissipation of the output transistor may result.

REVIEW

Questions

Q1. Is Q44 in Fig. 10-2 forward or reverse biased?

Q2. A vertical sync pulse is simply a series of horizontal sync pulses of longer duration. True or false?

Q3. All of the transistors in the vertical sweep circuits are operating as switches. True or false?

Q4. What symptoms would be apparent if capacitors C443 or C444 in Fig. 10-2 were low in value?

Q5. What would occur if Q44 in Fig. 10-2 shorted from base to emitter?

Q6. What is the function of the damper diode in solid-state horizontal-sweep circuits?

Q7. Is the horizontal-output transistor forward or reverse biased?

Q8. Horizontal oscillator failure can cause the output transistor to draw excessive current. Low-line-voltage operation should, for this reason, be avoided. True or false?

Q9. What will occur if C492 in Fig. 10-4 opens?

Q10. In Fig. 10-4, what will the collector voltage of Q49 be if a short is placed across the emitter-base terminals of Q48?

Answers

A1. Forward biased. The biasing path is through the emitter-base junction, R441, the transformer primary and diode, R442, and R44, the vertical hold control, to B+. The negative base voltage is a result of signal rectification.

A2. True. The horizontal pulses occurring during the vertical sync interval are about 3 times the width of a normal horizontal sync pulse.

A3. False. Only the oscillator transistor is switching, as is evidenced by the negative base voltage.

A4. Insufficient vertical scan would result. Q45 would not be driven strongly enough to produce a full raster.

A5. There would be no vertical deflection and the output transistor would saturate, drawing excessive current. A fuse or circuit breaker in the receiver would likely open depending on the dc resistance of the particular vertical output transformer. The emitter resistor R465 would overheat if the fuse or circuit breaker did not open.

A6. Prevent horizontal ringing after retrace is completed and also improve linearity and reduce the power dissipation of the output transistor.

A7. Reverse biased. No dc bias is applied which, in effect, reverse biases the transistor by about 0.6 volts (the diode drop).

A8. False. The output transistor is cut off when the oscillator fails.

A9. No raster. C492 is the ac return for the driver transistor. With C492 open, the signal currents through the primary of the driver transformer would be much too weak to drive the output stage.

A10. Close to zero. Q49 will saturate.

Index